"Crossing an impressive range of forms, industries, perspectives and practices, the contributions to this splendid collection are distinguished by a reparative impulse to locate, amid the wreckage signalled by the term 'Anthropocene,' the tools we need for constructing a co-habitable future. As these pieces show, cinematic forms can reveal and activate sustaining, Earth-focused forms of pedagogy, relationality, embodiment, perception, affect, spectatorship, ethics, and politics."

— Pansy Duncan (Massey University),
co-author of *Screening the Posthuman* (2023)

"Human understanding of the world around us has changed and it's time cinema studies caught up. This outstanding volume rises to that challenge with aplomb!"

— Stephen Rust (University of Oregon),
co-editor of *Ecocinema Theory and Practice* (2013)
and *Ecocinema Theory and Practice 2* (2023)

Cinema of/for the Anthropocene

Cinema of/for the Anthropocene sheds new light on the question of how films can allow us to resituate ourselves within what is known today as the Anthropocene. The authors address this question through a variety of disciplines and theoretical perspectives, from film and cultural studies, new materialisms, critical posthumanism and animal studies, critical race theory and Indigenous media studies, to gender and sexuality studies, with a primary focus on films produced in the United States and Canada.

The volume moves beyond the mere acknowledgment of the devastating damage inflicted during the Anthropocene to think about new ways of inhabiting the world through concepts such as affect, response-ability, and more-than-human kinship. The writers in this collection respond to its invitation by addressing a range of genres and modes, thus complicating the apocalyptic discourses which have typically been central to the studies on the Anthropocene: in addition to dystopian films, the volume discusses animated films, Hollywood biopics, climate change documentaries, experimental film, comedy, horror sci-fi, as well as disease thriller and survival film. Taken together, the chapters offer cross-disciplinary readings of the cinema of/for the Anthropocene, showing ways in which it can help us re-orient our thinking to make sense of the current age and address the planetary-scale environmental catastrophe.

This volume will appeal to researchers and students in film studies, cultural studies, and the burgeoning field of environmental humanities.

Katarzyna Paszkiewicz is an Associate Professor in English and Film Studies at the University of the Balearic Islands, Spain.

Andrea Ruthven is an Associate Professor in English at the University of the Balearic Islands, Spain.

Routledge Advances in Film Studies

Hybrid Documentary and Beyond
Rachel Landers

Epidemic Cinema
The Rise of a Genre
Julia Echeverría

The Disaster Film as Social Practice
Joseph Zornado and Sara Reilly

African Documentary Cinema
Alexie Tcheuyap

Eco-Teen Films
Robin L. Murray and Joseph K. Heumann

Centring Women in Bollywood Biopics
Empowerment and Agency in Contemporary Indian Cinema
Chandrava Chakravarty and Sneha Kar Chaudhuri

Cosmopolitan Aspirations in Contemporary Cinema
Edited by María del Mar Azcona, Julia Echeverría and Pablo Gómez-Muñoz

Cinema of/for the Anthropocene
Affect, Ecology, and More-Than-Human Kinship
Edited by Katarzyna Paszkiewicz and Andrea Ruthven

For more information about this series, please visit: Routledge Advances in Film Studies - Book Series - Routledge & CRC Press

Cinema of/for the Anthropocene

Affect, Ecology, and More-Than-Human Kinship

Edited by Katarzyna Paszkiewicz and Andrea Ruthven

Routledge
Taylor & Francis Group
LONDON AND NEW YORK

This publication was made possible by:
Research Project "Cinema and Environment: Affective Ecologies in the Anthropocene", funded by the Spanish Ministry of Science and Innovation (Grant PID2019-110068GA-I00 funded by MCIN/AEI/10.13039/501100011033).

First published 2025
by Routledge
4 Park Square, Milton Park, Abingdon, Oxon OX14 4RN

and by Routledge
605 Third Avenue, New York, NY 10158

Routledge is an imprint of the Taylor & Francis Group, an informa business

British Library Cataloguing-in-Publication Data
A catalogue record for this book is available from the British Library

ISBN: 978-1-032-74635-7 (hbk)
ISBN: 978-1-032-74182-6 (pbk)
ISBN: 978-1-003-46802-8 (ebk)

DOI: 10.4324/9781003468028

Typeset in Sabon
by SPi Technologies India Pvt Ltd (Straive)

Contents

Figures

Foreword

"Created to Dream": The Voices and Visions of *Cinema of/for the Anthropocene*

Salma Monani

In the cold opening of the animation short *Meneath: The Hidden Island of Ethics* (2021),[1] Métis filmmaker Terril Calder presents audiences with her protagonist, a big-eyed toddler with two high pigtails framing her baby face. Wearing nothing but a diaper and frilly socks and booties, Baby Girl scribbles with crayons at a desk, only to be startled as the image of the Native woman she is coloring over blinks at her before slipping out of the confines of the page. The camera takes Baby Girl's surprised perspective, an aerial shot looking straight down on the falling slender figure of the woman in her two-dimensional paper form. The woman, Nokomis (a revered person in Anishinaabe creation stories), is beautiful and unafraid even as she falls; her braids dance in the wind, and her jointed limbs, chest, and head imbue the white and yellow, red and black of the colored circular medicine wheel that forms her midriff.[2] As Nokomis falls, she looks straight up at Baby Girl, calling out "You were created to dreammmm," before landing with a splash and disappearing into water.

Barely has Baby Girl had a chance to absorb this surprise than a red-lipped, black-bearded but otherwise pale and loin-cloth-clad Jesus clears his throat and similarly escapes the confines of his page. The camera again takes Baby Girl's perspective as it frames his paper form hanging for a moment onto the page she holds in her dimpled hands. His words before he lets go are, "You were created out of original sin," and then, off he falls, to a similar fate, with a splash into the water.

As the camera lingers on the water surface, an island shaped as a turtle's shell appears with the two figures embraced on its back. Nokomis is curled protectively around Jesus, as the narrator notes, "sometimes they dance." Then, as the narrator continues, "sometimes they step on each other's toes," slick brown mud engulfs them both and the island. Everything disappears under a growth of moss as the title of the film, scribed with colorful plastic block alphabets that drop onto the rustling moss, closes out the cold opening.

The rest of *Meneath* unfurls with what is Calder's distinct oeuvre, a playful yet dark humor. The plot line involves Baby Girl wrestling with who she is as a Métis girl, and which stories to listen to—that of Christian impositions or that of her Anishinaabe roots—to ensure ethical obligations to self and others. *Meneath*'s attendance to such ethical considerations couched as they are in the gloriously irreverent, playful form of stop-motion animation has won it notable accolades. The Toronto International Film Festival (TIFF) named it as one of the best Canadian films of 2021 ("TIFF Canada's Top Ten"), and in 2022, it was nominated in the Best Animated Short category for the prestigious Canadian Screen Awards (Academy of Canadian Cinema and Television, 2022).

Given its content and form, *Meneath* would also win a spot in this anthology. Much like the films that follow in these pages, *Meneath* lends itself easily to what I and my colleagues, Sean Cubitt and Stephen Rust in our introduction to *Ecocinema Theory and Practice: Volume 2* (2023), describe as ecocinema studies' "radical rethinking of media and ecology engaged not only with intersectional questions of race, class, disability, gender, and sexual identities but also with the new configurations of social justice and the information economy" (2023, 4). Unearthing the complicated politics of a Native girl's "accommodation, misread as assimilation" (the opening words of the film), *Meneath* exposes issues of race and gender, and the brutality of patriarchal, settler-colonial impositions. In doing so, it also invites what Calder notes is the possibility of understanding and embracing "another set of ethics," than that of a patronizing Eurowestern persuasion. This other set of ethics, the film repeatedly reminds us, welcomes responsibility and responsiveness to human and posthuman relations of love, respect, and humility, i.e., to ecological kinship. *Meneath* is replete with animal iconography that animates the Seven Sacred Teachings of Anishinaabe that urge Baby Girl (and viewers) to a radical rethinking of how to care for oneself and others. Calder's filmmaking practices (as I have written elsewhere) are also attuned to rethinking how to produce cinema with an eye to social and ecological justice (Monani 2017).

The contributors to *Cinema of/for the Anthropocene* similarly urge such radical rethinking, bringing into this collection a critical commitment to expose cinema theory and practice to "other sets of ethics." Across its many chapters, much like *Meneath*, the anthology interrogates Eurowestern patriarchal systems. The opening chapter by William Brown sets the stage, exposing the Eurocentric (male) focus of traditional film history introductory courses, and arguing for dewesternizing the canon to help students grapple with important questions of which films get attention, and which get neglected. Incorporating cultural texts from the "margins," Brown argues for exposing students to "a different cinema." Refusing to satisfy various norms of the White gaze (such as those of certainty, closure, and separation from non-Whites as well as from nonhumans), such cinema, he convincingly argues, "might help us think a different world" from that which fuels our Anthropocene crises.

Contributors Libe García Zarranz, Missy Molloy, Andrea Ruthven, and Marta Segarra similarly help us "think a different world" through attention to films, like *Meneath*, that bring Indigenous communities' cultural sensibilities to light (Inuit, Māori, Cree/Métis, and Mapuche respectively). These chapters lend their voices to an emerging field of scholarship that elsewhere I describe as *d-ecocinema criticism*—a means of studying cinema as both ecological and decolonial in its onscreen messaging and offscreen labor and land practices (Monani 2024). These authors invite us to theorize how concepts central to cinema criticism (such as those of affect, temporality, and location) take on different meanings when read through "another set of ethics" than those of Eurowestern mainstream norms. For example, as Molloy suggests, what might it mean for (eco)cinema production to not "take" images of/from other-than-humans, but rather "make" cinema with the consent and collaboration of such others, as Māori filmmaker Nova Paul does in filming trees? Might such practices help bend the arc of filmmaking towards less exploitative endeavors than is often the case? Similarly, as she and Ruthven (and elsewhere I) argue by citing the work of Indigenous studies scholars such as Kyle Powys Whyte (Potawatomi), Zoe Todd (Métis), and Vanessa Watts (Mohawk), ecocinema scholars can "recalibrate their critical endeavors to explicitly acknowledge Indigenous contributions to our scholarly discourses." Opening cinema's critical endeavors to other ethics and perceptions can both redress past injustices of scholarly exclusion and enrich our understandings of media and ecology to invite more creative solutions to tackling the crises of the Anthropocene and otherwise.

Cinema of/for the Anthropocene shines a light not only on Indigenous thought but on other strands of intellectual thought that challenge narratives of Eurowestern exceptionalism. Chapters interrogate boundaries erected to denote what and who is understood as nature or culture (Luzón-Aguado's feminist reading), human or animal (Boczkowska's queer reading and Segarra's animist reading), ordinary or monstrous (Maleeva and Ungureanu's critical environmentalism reading, and Andermatt Conley's Deleuzean reading of "viral" networks). Many of the contributions encourage readings that ask us, as Brown writes in his chapter, "to live with the 'trouble' (Haraway 2016) of cinema's chaotic tendencies." Thus, Weik von Mossner's cognitive ecocritical reading reminds us that cinema's affective appeals (e.g., of hope and despair) empower or disempower action and agency depending on where the viewer stands, while Bergillos' chapter provokes readers to recognize how many of us, especially those from the Global North, live *in* media, while all of us live *in* climate change. Thus, he asks, how we might live in a world from which we cannot easily escape.

Presenting an array of intellectual traditions and a diversity of film genres, modes, and forms of production, this anthology is a cornucopia of thought-provoking inroads into the evolving landscape of ecocinema studies. Most hearteningly, *Cinema of/for the Anthropocene* joins a growing "third wave" of ecocinema scholarship by maintaining its focus on problematizing a long-held overemphasis on Eurowestern patriarchal structures that frame media and ecological studies (Monani and Rust 2023). Like with Baby Girl in *Meneath*, readers are welcomed to a journey to engage with alternative worldviews through which to "dream" and act on eco-social crisis and care.

Notes

1 Available at: https://www.nfb.ca/film/meneath-the-hidden-island-of-ethics/.
2 Nokomis is the Anishinaabemowin word for "my grandmother" ("Native Americans Not Just a Page in History", 2013). In various communities, including those of the Métis, she is also affiliated with creation stories of the world (Chilbow 2023).

Works Cited

Academy of Canadian Cinema and Television. 2022. "Canadian Screen Awards Nominees." *Academy.Ca* (blog). Accessed July 5, 2024. https://www.academy.ca/nominees/.

Chilbow, Susan. 2023. "Relationships and Responsibilities between Anishinaabek and Nokomis Giizis (Grandmother Moon) Inform N'bi (Water) Governance." *AlterNative: An International Journal of Indigenous Peoples* 19 (2): 283–92. https://doi.org/10.1177/11771801231173114.

Monani, Salma. 2017. "The Cosmological Liveliness of Terril Calder's *The Lodge*: Animating Our Relations and Unsettling Our Cinematic Spaces." *Studies in American Indian Literatures* 29 (4): 1–28.

Monani, Salma. 2024. *Indigenous Ecocinema: Decolonizing Media Environments*. Morgantown, WV: University of West Virginia, Salvaging the Anthropocene series.

Monani, Salma, and Stephen Rust. 2023. "'Dirtying' Ecocinema Studies: A Decade of Reflection." In *LIFE: A Transdisciplinary Inquiry*, edited by Janet Wasko and Jeremy Swartz, 270–86. Chicago, IL: University of Chicago Press.

"Native Americans Not Just a Page in History, Says Nokomis Learning Center." 2013. WMUK. September 16, 2013. https://www.wmuk.org/arts-more/2013-09-16/native-americans-not-just-a-page-in-history-says-nokomis-learning-center.

Rust, Stephen, Salma Monani, and Sean Cubitt, eds. 2023. *Ecocinema Theory and Practice 2*. New York: Routledge.

"TIFF Canada's Top Ten 2021." 2021. December 6, 2021. https://letterboxd.com/tiff_net/story/tiff-canadas-top-ten-2021/.

Acknowledgments

This book was supported by the research project "Cinema and Environment: Affective Ecologies in the Anthropocene," funded by the Spanish Ministry of Science and Innovation (Grant PID2019-110068GA-I00 funded by MCIN/AEI/10.13039/501100011033). Our preliminary results were presented at the international symposium "Cinema in the Anthropocene" (University of the Balearic Islands, 2022). The profound and engaging conversations held during this symposium lead to many of the brilliant contributions to the present volume. Warmest thanks to Suzanne Richardson, Stuti Goel and Suba Ramya Durairaj, and all the team at Routledge for so enthusiastically and carefully shepherding this volume, and to the anonymous reviewers for their vital and encouraging feedback. Thank you to Johns Hopkins University Press for permission to reprint Verena Conley's chapter, "Inhabiting a Viral Culture," which originally appeared in the journal *Substance* 51 (1), 2022. We would like to thank all the contributors to this collection for their excellent and engaging work, as well as numerous colleagues and interlocutors who, over the years, have helped us think more profoundly, and more capaciously, about cinema of/for the Anthropocene, and our own implications within it.

1 Thinking Cinema of/for the Anthropocene

An Introduction

Katarzyna Paszkiewicz

The Anthropocene: Too Early, Too Late?

It might seem paradoxical to write about cinema of/for the Anthropocene in the year in which the term 'Anthropocene' has officially been rejected, after more than two decades since its popularization. Employed by chemist Paul J. Crutzen and biologist Eugene F. Stoermer in 2000 to define a new planetary epoch characterized by human geological-scale impacts on earth's ecosystems, the term has stirred vibrant scholarly discussions on the socio-economic origins of the current environmental crises. While the concerns that the Anthropocene encapsulates remain pressing—anthropogenic climate change, accelerated species extinctions, rising oceans, and extreme weather events, as well as the disastrous consequences of voracious extractivism, among many others—in March 2024, after a series of votes, the International Union of Geological Sciences refused to declare an Anthropocene epoch.[1] Jan Zalasiewicz, a geologist and chair of the Anthropocene Working Group, has called for the result to be invalidated (Zhong 2024), while philosopher Timothy Morton (2024) has referred to the vote as "terrible blunder." "We need a name" for this epoch, he contends; "an epoch marked by the ravages of enslavement and settler colonialism, and of the oil-addicted automated version we call industrial capitalism" (2024).

Despite this recent 'Anthropocene controversy,' the concept has been in question for quite a while now, most prominently in the fields of the humanities and social sciences. Philosophers and cultural theorists have redefined and problematized the term in various ways, arguing that the Anthropocene, and the notion of a 'universal mankind' (*anthropos*) that underpins it, obscures existing power imbalances and varying degrees of responsibility and vulnerability, and it elides the agencies of non- or more-than-human entities, reinforcing human exceptionalism and the illusion of mastery over nature (Chakrabarty 2009; Nixon 2011; Alaimo 2016; Yusoff 2018; Oppermann 2018). Alternative terms have been proposed,[2] such as Capitalocene (Moore 2015), Plantationocene (Haraway 2016), or Wasteocene (Armiero 2021), which highlight the distinct spatio-temporalities of exploitation stemming from imperialism, colonialism, and capitalist

DOI: 10.4324/9781003468028-1

accumulation. Donna Haraway's (2016) work on the Chthulucene, in turn, has put forward a speculative, creative proposal centered on the "recuperation and recomposition" on an earth "full of refugees, human and not, without refuge" (100–101). Instead of relying solely on scientific reports or apocalyptic narratives of ecological disasters that often reinforce human exceptionalism, Haraway calls for new "stories (and theories)" (101) that can help us capture the complexity of our ecological entanglements, and embrace unexpected connections with the nonhuman, or even to dissolve the categorical divide between the human and the nonhuman. Meanwhile, critical race theorists and Indigenous scholars have challenged the novelty of these ideas, cautioning against the risk of obliterating the histories, experiences, and perspectives of enslaved and Indigenous peoples for whom the separation between the human and the nonhuman has always been a colonial imposition.

Regardless of the term or approach adopted, all these scholars express a shared commitment to pursuing less-destructive, habitable futures, and to questioning the binary oppositions that underlie the anthropocentric discourse, including human/nonhuman, life/matter, or nature/culture, thereby reframing the term Anthropocene itself. As anthropologist Sophie Chao notes in relation to these discussions:

> The greatest strength of the Anthropocene concept remains the way it helps unravel the presumed separation of humans from their environment, or of 'culture' from 'nature.' This in turn demands that humans take greater responsibility for the consequences of our unsustainable activities on earth. It further draws attention to the adverse consequences of anthropocentrism or human exceptionalism on planetary survival and wellbeing. In these respects, the Anthropocene is at once a characterization of the world we inhabit *and* a political call for change—an 'action-concept' *par excellence.*
>
> (2024)

In dialogue with these debates, film studies has increasingly focused on understanding cinema as a means of exploring realms beyond the anthropocentric. Over the past decade, we have witnessed an increased proliferation of studies on ecocinema, as well as, more recently, on the posthuman/nonhuman on screen. This can be exemplified by the publication of two edited collections on *Ecocinema Theory and Practice* edited by Stephen Rust, Salma Monani, and Sean Cubitt—the first in 2013 and the second in 2023—which indicates a consolidation of ecocinema studies in the last ten years.[3] The ecological turn in film studies has led to a systematic exploration of different forms of ecocinema, which challenge prevailing viewing habits and encourage a shift from "a narrow anthropocentric worldview" to an "earth-centered" or "ecocentric" perspective (Willoquet-Maricondi 2010, 46). Cinema's non-hierarchical, and thus, potentially non-anthropocentric opening to the world, and its capacity to foster affective attunement with the nonhuman, has been central in ecologically informed approaches, which were initially applied to experimental and documentary films.[4] However, the burgeoning scholarship on climate fiction (or cli-fi), and eco-dystopian and eco-utopian films more broadly, demonstrates that popular cinema and Hollywood genres are also well suited to foster ecopolitics, in both cognitive and affective terms (see Alexa Weik von Mossner's and Julia Leyda's numerous works on the topic).[5] Further, the recent investigations into the posthuman/nonhuman in contemporary film, including Chelsea Birks's 2021 *Limit Cinema*, Missy Molloy, Pansy Duncan, and Claire Henry's *Screening the Pothuman* (2023), and the special issue on "Film and the Nonhuman" (2024), edited by Cristóbal Escobar and Barbara Creed for *Senses of Cinema*, demonstrate the fruitfulness of examining new models of non-anthropocentric subjectivity across a variety of generic, national, and industrial contexts. The growing field of environmental film studies has encompassed the consideration of the materialities of cinema and its infrastructures, such as the film industry's reliance on natural resources—building upon earlier studies by Bozak (2011) and LeMenager (2014) on this subject—or the carbon footprint of streaming platforms more generally, which is how most films are viewed nowadays (Marks 2020). It has also increasingly addressed the questions around eco-justice, and embraced postcolonial, decolonial, and/or intersectional perspectives, including Indigenous epistemologies and Black media studies (Rust, Monani, Cubitt 2023; Brown 2023; Leyda 2023; Monani 2024; Telles 2024, among others), which help us to rethink the ways in which film studies in general, and ecocinema and Anthropocene film studies in particular, can be dewesternized.

The growing interest in these issues within film studies is also attested to by the proliferation of concepts: in addition to ecocinema and cli-fi, these include cinematic weather (McKim 2015) and elemental cinema (de Luca and Mroz 2023); climate trauma (Kaplan 2016) and eco-trauma cinema (Narine 2018); Anthropocenema (Kara 2016), and more recently, Haraway-inspired Chthulucinema (Uy and Brown 2020). Rather than adding to these concepts, the present volume proposes the development of conceptual frameworks that would allow for thinking about films of/for the Anthropocene from an intersectional and cross-disciplinary perspective, while maintaining a focus on the questions of gender, sexuality, Blackness, Indigeneity, and animality, among others. In doing so, it builds on a number of monographs published over the last few years, beginning with the groundbreaking *Inhospitable World: Cinema in the Time of the Anthropocene* (2018) by Jennifer Fay. Fay's theoretical argument revolves around cinema's ambition to create artificial weather and "inhospitable worlds" throughout film history, from Buster Keaton to film noir and films of Antarctic exploration. Engaging in dialogue with this book, William Brown (2023) proposes the concept of the Black Chthulucene to unpack the white supremacy that often underpins western theorizations on the cinema in the Anthropocene. Arguing that ecology, just like the Anthropocene, "emerges as a term that is too focused on whiteness/the white *oikos*," he adds: "The construction of the white *oikos* on a planet that whites otherwise render 'inhospitable' is indeed the construction of *economy*, literally the law of the home/house, or the house as law, or capital itself" (136). Julia Leyda's *Anthroposcreens* (2023), in turn, offers a reading of "the climate unconscious" in relation to Black American and Norwegian film and television as a counterpoint to Anglo-American cli-fi. As she observes, mainstream climate fiction all too often "repeats Hollywood conventions that dramatize the survival of White patriarchal reproductive futurism embodied by the nuclear family" (2023). These reflections are highly relevant for the present collection, as many of the essays (including Brown's contribution to our volume) offer tools for thinking about cinema and the Anthropocene beyond universal mankind, heteronormativity, and/or whiteness.

Beyond the Apocalypse: Affect, Ecology, and More-Than-Human Kinship

Donna Haraway (2016, 2) has famously argued that we need to learn "to stay with the trouble of living and dying in response-ability on a damaged Earth." However, she also urges us to resist apocalyptic discourses, or what she calls, in her dialogue with Marta Segarra, "the apocalypse of the *jouissance* of total destruction" (Haraway and Segarra 2019, 90), because it tends to paralyze us. Acknowledging the scale of the damage is important, but equally important is the partial rehabilitation of the earth in ways that would make sense both to human and more-than-human lives.

Indeed, with some notable exceptions,[6] disaster movies and (post)apocalyptic scenarios of collapse that dramatize 'the end of the world' have been overrepresented in scholarship on the Anthropocene cinema. Until the present moment, most of these reflections have focused on sci-fi, dystopian fiction or, more recently, ecohorror and the Gothic (White 2019; Pesses 2021; Edwards et al. 2022; Tidwell and Soles 2021; Troon 2024), highlighting apocalyptic discourses on the Anthropocene, rather than the focus on (partial) rehabilitation and modes of rebuilding. The present collection aims to redress this imbalance, by engaging with a range of genres and industrial contexts, including Hollywood biopics, horror sci-fi, climate change documentaries, Indigenous animated films, experimental cinema, comedy, as well as the disease thriller and survival films. In doing so, it explores a variety of topics which are relevant for the discussions of the Anthropocene, such as the questions of exposure and response-ability; climate activism; Indigenous filmmaking practice and more-than-human agency; Blackness and the Anthropocene; animality and human-animal hybridity; posthuman and more-than-human kinship; media infrastructures in the Anthropocene; global capitalism, viral cultures, and the ethics of care. It reflects on how cinema not only gives an aesthetic and affective form to the Anthropocene, but also how it can allow us to rethink our relationship with the environment and its more-than-human inhabitants. These questions are framed within the recent scholarship on the ecological in the humanities, with a special emphasis on the lines of research emerging from within the environmental humanities, including new materialisms and critical posthumanism, as well as decolonial and Indigenous epistemologies, which are crucial in redirecting current debates on cinema and the environment. The contributors to this book work across various disciplines, including film and media studies, cultural studies, Blackness and Indigeneity studies, ecofeminism, queer ecofeminism and trans ecology, affect studies, as well as critical animal studies. The films addressed in these pages are not the only possible candidates for the study of cinema in the Anthropocene (for instance, ecocinematic filmmaking in South Asia or Latin America also invites discussion). The selection of texts for the present study has been limited, to an extent, by the scope of our research project from which the volume has emerged and to our scholarly expertise.[7] We are convinced that the inclusion of perspectives from beyond the Anglophone (or, indeed, western) world is certainly needed to decenter the historically contingent Euro-American criticism. Further, while most of the films analyzed in this book have been produced in the 21st century, it is important to note that cinema has long been imagining or even responding to the Anthropocene. This is acknowledged by William Brown's study of silent-era films included in the volume.

Several of the essays in this collection articulate the discussions on the Anthropocene in relation to the questions of affect, embodiment, and the senses (García Zarranz, Weik von Mossner, Molloy, Boczkowska, Conley). Despite the frequent opacity of the much-debated term 'affect,' the chapters in this book suggest that it can prove useful for examining the collective sensorium and responses to the current environmental crisis, which are formalized in and shaped by contemporary cinema. Further, a consideration of multiple senses and visualities, that foster a bodily, and not purely visual, reciprocity with the world, is particularly relevant for thinking about cinema of/for the Anthropocene, especially when combined with theorizing our affective, in addition to cognitive, engagements with films. As Alexa Weik von Mossner's pioneering work has demonstrated (2014; 2017), the disciplinary intersection of film studies, affect studies, and ecocritical scholarship can be particularly fruitful when addressing how films can "invite us to care for human and nonhuman others who are put at risk" (2017, 4).[8] Although Weik von Mossner's conceptualization of affect is connected to cognitive narratology and neuroscience, aligning with the cognitive ecocritical framework she has proposed, her generative term "affective ecologies" can be broadened to include various other ecocritical approaches.[9] Julia Leyda's work (2023) on "the climate unconscious" also unpacks aesthetic and affective responses to contemporary environmental crises, adopting Raymond Williams's notion of "structures of feeling," a term that suggests that affects, emotions, and feelings are not solely individual experiences but are intricately linked to their social and historical context. It is this enlarged, capacious understanding of affect that guides the present collection, as the volume aims to illustrate how different cinematic forms can activate new perceptions and reciprocity with the world, creating different sorts of cinematic identification, affinities, or embodiment. More than addressing compassion or empathetic relation—understood as an individual psychological state that often presupposes hierarchy—the aspiration of this volume is to come up with inclusive, hybrid, and open-ended approaches that would fracture the paternalistic assumptions implied in traditional understandings of 'taking care of' the environment.[10]

Crucially, the works analyzed in these pages complicate the very notion of 'environment,' as well as those of 'nature' and 'agency' (Weik von Mossner, Luzón-Aguado, Bergillos, Ruthven). Examining the idea of cinema of/for the Anthropocene in dialogue with new understandings of the relationship between humans and their more-than-human environments helps to question such a distinction (Barad 2007; Alaimo 2010; Bennett 2010; Braidotti 2013; Whyte 2017; Simmons 2019). Historically, studies from an ecocritical perspective have primarily focused on films considered 'environmentalist,' especially documentaries or Hollywood films that idealize nature and its defenders (such as *Gorillas in the Mist* or *Erin Brockovich*, to give some well-known examples), often perpetuating, rather than destabilizing, the anthropocentric gaze. Yet, within the context of the Anthropocene, a term that, as previously noted, itself carries several contradictions, the concept of the 'environment' has gained additional layers of complexity. Media scholar Sean Cubitt (2019) famously argues "against environmentalism," joining others in radically rethinking ecological models based on the traditional notions of 'environment' or, more commonly, 'nature' (see, for instance, Timothy Morton's oft-quoted argument for "ecology without Nature"). The concept of 'ecology' is equally problematic as it "presumes there are things to connect; but the truth is that there are only connections, and the connections produce the 'things'" (Cubitt 2019). These re-evaluations often engage in dialogue with perspectives such as critical posthumanism, animal studies, and new materialisms, which frequently emerge from feminist, postcolonial/decolonial, and queer frameworks. Consistently demonstrating a keen interest in the materiality of difference, these frameworks reject the supposedly universal humanism that underlies discussions of the Anthropocene (Braidotti 2013; Alaimo 2016; Morton 2009; Ghosh 2016; Ferdinand 2019).

Jane Bennett's (2010) influential argument for a "more distributive agency" seeks to further dissolve the ontological binaries of "life/matter, human/animal, will/determination, and organic/inorganic," highlighting the ethical and political implications of this endeavor (ix-x). An increased attentiveness to matter and its powers does not necessarily solve the issue of exploitation or oppression, but it fosters a greater recognition of how all bodies are "kin in the sense of [being] inextricably enmeshed in a dense network of relationships" (13). Bennett's "vital materialist theory of democracy" (108) is of particular interest to film scholars who have now started to examine how cinema attends to the interconnections between species without reinforcing hierarchical distinctions (see, for instance, McMahon 2014). Such retraining of attention and the focus on cross-species relationality undermine the dominant discourses on the Anthropocene for, as Alaimo argues, "whatever the 'anthro' of the 'anthropocene' was, is, or will be, the anthropocene must be thought with the multitude of creatures" (2016, 143). This is why this collection places a particular emphasis on the questions of cinematic animality (Segarra, Boczkowska), as well as instances of posthuman or more-than-human kinship (García Zarranz, Molloy, Maleeva and Ungureanu), extending the notion of kinship not only beyond patriarchal heteronormativity, but also beyond the species divides.[11]

Last but not least, the reflections on the cinema in the Anthropocene, as well as more-than-human kinship, need to be placed in conversation with Indigenous and Black thought as a means of articulating a position from which to redress the overemphasis on Euro-western frameworks. As geographer Kathryn Yusoff pointed out, "the Anthropocene might seem to offer a dystopic future that laments the end of the world, but imperialism and ongoing (settler) colonialism have been ending worlds for as long as they have been in existence" (2018, xiii). Several chapters of this volume (Brown, García Zarranz, Molloy, Ruthven) question anthropocentric narratives that often obscure the histories of colonial expansion and massive exploitation that continue to the present day, while engaging in a fruitful dialogue with a range of approaches for understanding more-than-human relations, agency, and reciprocity, beyond neoliberal, individualist identity formation (Watts 2013; Todd 2016; Cariou 2020; Monani 2024). These chapters contribute to both decolonizing ecological approaches to cinema and highlighting our own positionality as scholars working within particular cultural and socio-economic contexts: Europe, the US, and Canada. As Brown states in the present volume, the endeavor to dewesternize the texts we choose to watch, analyze or teach "is also an endeavor to decolonize the western perspective, leading hopefully to a sense that western filmmaking presents to us not just stories but also a worldview … that are not timeless and universal, but rather historically contingent." Ethical relations with the more-than-human world and recognition of agency and liveliness in the more-than-human other, which form the cornerstone of many Indigenous epistemologies, illuminate our own understandings of climate justice and sustainable futures.

Films of/for the Anthropocene

One of the key questions of the present volume is how cinema can cultivate response-ability, which entails avoiding an "unprecedented looking away" (Haraway 2016, 35). The first section, "Affect, Ecology, and Pedagogies of Worldly Reciprocity," builds on what Lauren Berlant describes in *Cruel Optimism* as "our sense of reciprocity with the world" (2011, 52). For Berlant, aesthetics is crucial in apprehending the conditions of the present, as it allows us to "rehabituate our sensorium by taking in new material"

(12). As John Landreville has shown, these ideas turn out to be particularly meaningful when thinking about ecocinema, which can be defined as a "pedagogy of worldly reciprocity", that is, "the site where humanity's relationship to worldhood is re-learned" (2019, 4). Thus, the chapters in this section open up questions about how different film forms can mobilize the retraining of attention (MacDonald 2013), while also focusing on the interaction between aesthetic, affective, and ideological meanings involved in cinematic renderings of the more-than-human environment.

William Brown's opening contribution shows how, since its inception, cinema has always been imagining the end of the world, thus revealing cultural anxieties about the Anthropocene. In his comparative reading of the silent-era films *The Navigator* (Buster Keaton and Donald Crisp, United States, 1924) and *Kaliya Mardan* (D.G. Phalke, India, 1919), Brown argues that while the former film points to how the animal and the Black/Indigenous other need to be exterminated to make room for the heteronormative bourgeois white couple, the latter "teaches us to be mindful of (rather than threatened by) the outside." Through this discussion, the chapter calls for a new pedagogy that would challenge the canon of Euro-American cinema by dewesternizing our syllabi, while offering a reframing of film history through the lens of Haraway's concept of the Chthulucene. Such rethinking has significant implications for contemporary cinema of/for the Anthropocene, as it points not only to its role in the "Americanization" of the world and the instigation of the exploitative practices and attitudes that are based on a sense of separation from the lifeforms we are entangled with, but also to cinema's potential for helping us to think a different world through the aesthetics (and ethics) of relationality that entangles the audience in the process.

Also highlighting the relational aspects of film viewing, Libe García Zarranz's chapter considers Bessie P. Dernikos's (2018) pedagogies of exposure as a means through which to examine Jonathan Wright's animated film *The Bear Facts* (Nunavut Animation Lab, 2010), which reimagines first contact between a colonial explorer and an Inuit hunter. Drawing on affect studies, Indigenous knowledges, and feminist and critical race theories, García Zarranz reads the film in connection with Aphton Corbin's *Twenty Something* (Pixar, 2021) to argue that cinema can not only expose but also heal traumatic wounds through what she conceptualizes as willful aesthetics. The concept refers to both film techniques that mobilize sensorial and affective realms (the use of light and sound, 2D animation, watercolor landscaping), as well as the characterization of "insurgent" human and more-than-human subjects that centers Indigenous knowledges and Black kinship. In so doing, the chapter raises wider questions about how films can provoke viewers to respond affectively, ethically, and aesthetically to the current environmental crisis and how this crisis necessarily intersects with other forms of exploitation, such as settler colonialism and anti-Black necropolitical forces.

Cinema's aesthetic and affective strategies as drivers of transformative action are also central in Alexa Weik von Mossner's chapter on Jim Rakete's

Now (Germany, 2021) and Dave Davis and Mary Grandelis's *Being the Change* (United States, 2018), two climate change documentaries which avoid the mere exposition of scientific facts or the use of apocalyptic imagery. Situating the films within the wider debate on hope and agency in diverse fields, including philosophy, psychology, sociology, environmental history, literary studies, and pedagogy, Weik von Mossner explores the affirmative potential of cinema and its role in both the creation and sharing of social dreaming. Engaging in dialogue with Ernst Bloch's (1986) *The Principle of Hope*, and the concept of "the courage of despair," the chapter demonstrates how hopeful climate storytelling in nonfiction films can make us both *think* and *feel*. Weik von Mossner contends that while these environmental documentaries might have modest viewership in comparison with the fictional envisioning of more sustainable futures, limited mostly to audiences in the industrialized Global North, they can nonetheless serve as catalysts for change and lead to broader discussions about intergenerational climate justice.

Virginia Luzón-Aguado's chapter similarly raises questions about affirmative aesthetics and agency, but it focuses attention on the survival-in-the-wilderness film, a subgenre of the action-adventure narrative. Building on both genre theory and ecofeminism, Luzón-Aguado offers a reading of *Wild* (Jean-Marc Vallée, United States, 2014) and *Tracks* (John Curran, Australia, 2013), two films that thematize women trekking alone on the Pacific Crest Trail in the US and the deserts of Western Australia, respectively. The chapter points to the reasons why western women have been conventionally excluded from "walking in the wild" in the context of mainstream cinema, while scrutinizing the discursive connection between femininity and the natural world. Like the previous chapters, it also challenges the imaginary of the fertile "Mother Nature" as a resource to be tamed and domesticated, as well as the thoroughly masculinized, capitalist narratives about possessing and exploiting the land.

The section closes with Ignacio Bergillos's reflection on the role of media in how we perceive and inhabit the so-called environment, which can be considered yet another facet of the pedagogy of cinema of/for the Anthropocene, as it takes into consideration the material infrastructures of media, cinema included, and their contribution to the devastation of the planet. Peter Weir's satirical comedy *The Truman Show* (United States, 1998) serves as a point of departure to interrogate our ambivalent relationship with the Anthropocene as experienced in the Global North: the feeling of both powerlessness and the urgency for action. Specifically, the chapter expands on Mark Deuze's (2012) idea of the pervasiveness of media—how contemporary societies do not simply *live with* media but rather *live in media*—to illustrate the implications for thinking about environment and media as closely interrelated and similarly unpredictable. The chapter further proposes a nuanced conceptual framework to better understand human agency in "an age of ubiquitous media and climate change," posing the following question: what is there to do (in a media life/in the Anthropocene) once we know that there is no exit (like in

Truman's case) or no alternative planet (in our case)? By examining how cinema can foster worldly reciprocity, all of the chapters in this section offer approaches to films that not only acknowledge the damage of the Anthropocene, but also propose new understandings of cinema and ecology, which are contingent on cultural and historical contexts in ways that demand constant rethinking.

In view of the problematic status of the concept 'Anthropocene,' which more often than not reinforces the binaries underlying the anthropocentric view (such as human/animal or nature/culture), one of the main aims of our collection is to consider how cinema can challenge the dichotomous thinking that has kept us apart from the environment and contributed to racism, speciesism, and other exclusions. These binaries therefore come under scrutiny in Part 2, entitled "More-Than-Human Kinship, Hybridity and Monstrous Alliances," which unites essays focusing on the co-habitation with human and more-than-human organisms through an emphasis on shared relationalities and co-constitution, often in conversation with non-western frameworks (especially considering that 'kinship' resonates so strongly with many Indigenous worldviews, both historically and in contemporary Indigenous studies).[12]

The section opens with Missy Molloy's chapter on the cinema of Nova Paul, which reframes the relationship between humans and their more-than-human environments through Māori notions of time and kinship. While considering Nova Paul's practice as an instance of new materialist filmmaking, especially in the way it "involves collaboration with the material world," Molloy grapples with the Indigenous-centered critique of both the discourse on the Anthropocene and new materialisms. Such a critique, she contends, is remarkably consistent among Indigenous scholars across the globe (including the US and Canada), and she wonders why non-Indigenous scholars "seem less cognizant of (or engaged with) this work." The chapter urges western scholars to take up these "kindred theories," not only to acknowledge Indigenous contributions to contemporary ecological thought, but also to complicate and expand upon our understandings of the dynamic between humans and the nonhuman world.

The next chapter, by Andrea Ruthven, also offers a discussion of Indigenous filmmaking and eco-social resilience, focusing on the context of settler colonialism in Canada. The chapter draws on Indigenous and decolonial thought to critically explore the concepts of "terristories" (Cariou 2020) and "Place-Thought" (Watts 2013) in ways that challenge western post-apocalyptic narratives of the Anthropocene. In her discussion of Cree/Métis filmmaker Danis Goulet's short film *Wakening* (2013) and Cree/Michif filmmaker Amanda Strong's stop-motion film *Biidaaban (The Dawn Comes)* (2018), Ruthven relies on Indigenous epistemologies that understand the land as inherently agential and animate, especially in the ways in which it communicates through stories—even though such communication is often "muted within settler-colonial habitats." Further, the chapter shows how, by placing

Indigenous storytelling within colonial urban settlements, these films draw from a long tradition of site-specific Indigenous epistemologies that challenge a nature/culture divide which underpins many of the conventional readings of non-western epistemologies and artistic practice.

There follows Marta Segarra's chapter that opens with a theoretical focus on animism and human–animal relationships—a highly pertinent area of study, given that we now live in the age of the "sixth extinction" (Kolbert 2014), that is, a loss of biodiversity of breath-taking magnitude. The chapter addresses the (only apparently) paradoxical relationship between the western belief in human exceptionality and our dreams of human–animal hybridization as reflected in cinema and other arts. After considering how such hybridity is often represented as abnormal and employed to reaffirm the boundaries of humanity, this essay moves on to offer a comparative reading of four films from different geographical areas, thus extending the growing body of Anglophone work on cinematic animality to European, world cinema, and Indigenous contexts: James Marsh's *Project Nim* (United Kingdom/United States, 2011), Apichatpong Weerasethakul's *Tropical Malady* (Germany/France/Italy/Thailand, 2004), Francisco Huichaqueo's *Gente pájaro. Che Uñum* (Chile, 2007), and Gerard Ortín's *Agrilogistics* (Spain/United Kingdom, 2022). All of these films illustrate the porosity of the previously assumed limits between human animals, other-than-human animals, and the supposedly inanimate world of objects. Further, Segarra's thought-provoking conceptualization of "shamanic cinema," in connection with Hélène Cixous's "animal writing," opens up questions about cinema itself being "like an animal."

Kornelia Boczkowska's chapter further develops the theme of human and nonhuman animals on screen, focusing specifically on human–horse relations as represented in two experimental films: Barbara Hammer's *A Horse Is Not a Metaphor* (United States, 2008) and Ann Oren's *Passage* (Germany, 2020). The films under discussion challenge the hierarchical structures of domination and mastery associated with "looking at animals" (Berger 1980) and blur the animal–human boundaries, avoiding the anthropomorphization that is attached to most cinematic representations of the nonhuman to date. Drawing on critical animal studies, queer ecofeminism, and trans ecology, the chapter shows how the films' highly stylized mise-en-scène and editing choices contribute to disrupting a gendered binary between nature and culture, while fostering a queer fluidity of human–horse hybrids on screen.

The next chapter develops yet another facet of human–nonhuman relationality and the Anthropocenic condition by moving to a consideration of the rarely examined issue of genetic hybridization.[13] In particular, it focuses on the sci-fi horror film *Annihilation* (Alex Garland, United Kingdom/United States, 2018) and Jeff VanderMeer's homonymous novel (2014), on which the film was based. Building upon critical environmentalism that avoids the utopia/dystopia binarism in reframing human–nature relationships, Galyna Maleeva and Camil Ungureanu argue that, while both texts challenge traditional subject–object divides, the representation of transspecies relationships and wondrous alterity operate differently in each of them. While VanderMeer's source narrative dramatizes the move from the protagonist's autonomous subjectivity to a transformative fusion with the "alien nature," which questions the borders between human and nonhuman entities, Garland's adaptation remains, in part, confined to an anthropocentric scenario, thus limits rather than expands on the radical possibilities of posthuman kinships. Yet, at the same time, both texts envision intriguing alternatives for the anthropocentric discourse that sees nature as an object to be mastered by humans, gesturing toward monstrous alliances, hybridity, and more-than-human agency, or even, as Jane Bennett dubs it, "vibrant matter" (Bennett 2010).

The book closes with Verena Conley's meditation on what she calls "viral cultures," and how they are illuminated by Steven Soderbergh's prescient film *Contagion* (United States, 2011), and the recent COVID-19 pandemic—which, as has been widely demonstrated, is closely interrelated with the exploitation of resources and the global impact of advanced capitalism in the Anthropocene. Like the previous chapter in this section, it rethinks the monstrous (or "evil") nonhuman other, in this case viruses, to consider how we are all enmeshed with animals, plants, and inorganic matter. Indeed, telling stories of how deadly viruses infiltrate and kill humans elides the fact that they are vital for all living creatures. This approach turns on its head the common idea of "bad viruses" against which a war must be waged and focuses instead on issues of care. It also overthrows the narratives of control over nature, showing how humans are not—and have never been—as in control as previously assumed. Finally, the chapter brings readers back to the first section of the book by showing how Soderbergh's film helps us think about the concept of ethology in relation to a circulation of affects, as proposed by Deleuze in his reading of Spinoza's *Ethics*. For the latter philosopher, ethics does not involve moralizing, but rather, ways of being and inhabiting the world, which is felt in and through our bodies, and not simply observed at a distance. Conley's chapter shows how *Contagion*'s cinematic aesthetic both sheds light on and obscures this message. The chapter thus closes our collection with an ethically grounded focus, by tracing a strong line between film aesthetics and the call to alter our modes of perceiving the Anthropocene. This is vital because, as media ecocritics argue (Cubitt 2016), it is increasingly through media, cinema included, that the world is sensed. Taken together, the essays in this collection suggest new ways of cinematic seeing and feeling (for) the Anthropocene, while critically grappling with the tensions inherent in the concept. They shed much-needed light on more-than-human co-existence and co-becoming, imagining less violent and more habitable futures.

Notes

1 The proposal that was voted against aimed to mark the year 1952 as the beginning of the Anthropocene to reflect humanity's impact on the planet (when radioactive plutonium coming from hydrogen bombs appeared in the sediment of

Crawford Lake in Canada). According to *The Guardian* article that first published the results of the votes, "those objecting noted a much longer history of human impacts on Earth, including the dawn of agriculture and the industrial revolution, and unease about including a new unit in the geological timescale with a span of less than a single human lifetime" (Carrington 2024). The issue of timescale has been long debated. Crutzen originally considered the Industrial Revolution (1800s) as the beginning of the Anthropocene, while others pointed to the colonization of the Americas (1492), the first testing of the atomic bomb (1945) or the Great Acceleration (1950s).

2 Like the Anthropocene, these terms have also been associated with certain omissions or conceptual limitations (see Chao 2024). However, as Chao proposes, "the various labels need not be approached as exhaustive or mutually exclusive" (2024).

3 See also the recent publication of *Routledge Handbook of Ecocinema Studies* (López et al. 2024).

4 See, for instance, Scott MacDonald's (2013) influential work on ecocinema, Anat Pick's engagement with zoomorphic realism (2011) or James Leo Cahill's (2019) study of zoological surrealism in the cinema of Jean Painlevé.

5 See Leikam and Leyda (2017), and Weik von Mossner (2017), among other works.

6 In addition to Fay (2018), Brown (2023) and Leyda (2023), Mark Bould's 2021 short monograph *The Anthropocene Unconscious: Climate Catastrophe Culture* explores how anthropogenic climate emergency manifests itself in texts that are not at first readable as texts about the Anthropocene. In contrast to our collection, it addresses not only film, but also television and literature, and its main focus is on climate change. Molloy, Duncan, and Henry's *Screening the Posthuman* (2023) also makes a case for expanding the scope of genres when addressing posthuman subjectivities and the more-than-human relationality.

7 This book stems from the research project "Cinema and Environment: Affective Ecologies in the Anthropocene," funded by the Spanish Ministry of Science and Innovation (Grant PID2019-110068GA-I00). Our research team encompassed scholars specializing in US and Canadian culture, as well as in ecocriticism and affect studies: Verena Conley (Harvard University), Libe García Zarranz (Norwegian University of Science and Technology), Katarzyna Paszkiewicz (Principal Investigator, University of the Balearic Islands), Andrea Ruthven (University of the Balearic Islands), and Alexa Weik von Mossner (University of Klagenfurt). See: https://cinemaenvironment.uib.eu/.

8 See also Adrian J. Ivakhiv's (2013) monograph *Ecologies of the Moving Image: Cinema, Affect, Nature*, which offers an ecophilosophical account of the moving image across a variety of genres, building on process-relational philosophers such as Whitehead, Peirce and Deleuze; as well as Salma Monani's (2024) work on Indigenous affects in the context of what she dubs "d-ecocinema criticism."

9 For a more detailed consideration of affect theory in relation to ecocinema, see Paszkiewcz (2021).

10 Although not extensively examined in the present collection, it is worth mentioning the studies on "negative" or "dissident" affects, as well as "bad environmentalism," as explored by queer ecology (Morton 2010; Seymour 2018, among others), which help to counter a utopian, nostalgic aspiration of 'connecting' with the environment, typical of the dominant environmental politics. Queer ecology also sheds light on 'uninvited' kinships with more-than-human others, including viruses, bacteria, parasites, and so on. The latter are addressed in Verena Conley's chapter included in this collection.

11 See also our previous work on "posthuman kinship," which provides a brief overview of the theoretical underpinnings of the concept (Paszkiewicz, Ruthven,

Segarra 2021). Some writers in this collection adopt the notion of "more-than-human kinship," reflecting their emphasis on relational ethics, ecological interconnectedness and embeddedness of all life forms in the material world (see, for instance, Stacy Alaimo's use of the term "more-than-human").

12 Indeed, the idea of posthuman or more-than-human kinship is central to many non-western understandings of humanity as always "constituted in inextricable relations with the nonhuman world" (Bignall and Rigney 2019, 159). Writing on Indigeneity, posthumanism and "being-more-than-human," Simone Bignall and Daryle Rigney (2016, with Steve Hemming; 2019) point out that such philosophies—ranging from pantheism, genealogical constructivism to expressivism—have long rejected anthropocentrism and human exceptionalism.

13 However, there are some notable exceptions, for instance, Pansy Duncan's (2023) recent conceptualization of the body as cell in Garland's *Annihilation*, as part of the broader discussion of what she dubs the "eco-material posthumanism."

Works Cited

Alaimo, Stacy. 2010. *Bodily Natures: Science, Environment, and the Material Self.* Bloomington: Indiana University Press.

Alaimo, Stacy. 2016. *Exposed: Environmental Politics and Pleasures in Posthuman Times.* Minneapolis: University of Minnesota Press.

Armiero, Marco. 2021. *Wasteocene: Stories from the Global Dump.* Cambridge: Cambridge University Press.

Barad, Karen. 2007. *Meeting the Universe Halfway: Quantum Physics and the Entanglement of Matter and Meaning.* Durham: Duke University Press.

Bennett, Jane. 2010. *Vibrant Matter: A Political Ecology of Things.* Durham: Duke University Press.

Berger, John. 1980. "Why Look at Animals?" In *About Looking*, 3–28. London: Bloomsbury.

Berlant, Lauren. 2011. *Cruel Optimism.* Durham: Duke University Press.

Bignall, Simone, and Daryle Rigney. 2019. "Indigeneity, Posthumanism and Nomad Thought: Transforming Colonial Ecologies." In *Posthuman Ecologies: Complexity and Process after Deleuze*, edited by Rosi Braidotti and Simone Bignall, 159–83. London and New York: Rowman and Littlefield International.

Bignall, Simone, Steve Hemming, and Daryle Rigney. 2016. "Three Ecosophies for the Anthropocene: Environmental Governance, Continental Posthumanism and Indigenous Expressivism." *Deleuze Studies* 10 (4): 455–78.

Birks, Chelsea. 2021. *Limit Cinema: Transgression and the Nonhuman in Contemporary Global Film.* London: Bloomsbury Academic.

Bloch, Ernst. 1986. *The Principle of Hope.* Cambridge, Mass.: MIT Press.

Bould, Mark. 2021. *The Anthropocene Unconscious: Climate Catastrophe Culture.* New York: Verso Books.

Bozak, Nadia. 2011. *The Cinematic Footprint: Lights, Camera, Natural Resources.* New Brunswick: Rutgers University Press.

Braidotti, Rosi. 2013. *The Posthuman.* Cambridge: Polity Press.

Brown, William. 2023. *Navigating from the White Anthropocene to the Black Chthulucene.* Hampshire: Zero Books.

Cahill, James Leo. 2019. *Zoological Surrealism: The Nonhuman Cinema of Jean Painlevé.* Minneapolis: University of Minnesota Press.

Cariou, Warren. 2020. "Terristory: land and language in the Indigenous short Story – Oral and Written." *Commonwealth Essays and Studies* 42 (2): 1–9.

Carrington, Damian. 2024. "Geologists reject declaration of Anthropocene epoch." *The Guardian*, March 22, 2024.

Chakrabarty, Dipesh. 2009. "The Climate of History: Four Theses." *Critical Inquiry* 35 (2): 197–222.
Chao, Sophie. 2024. "Beyond the Anthropocene: We must rethink our epoch." *IAI News*, April 8, 2024. https://iai.tv/articles/beyond-the-anthropocene-we-must-rethink-our-epoch-auid-2803.
Crutzen, Paul J., and Eugene F. Stoermer. 2000. "The 'Anthropocene.'" *Global Change Newsletter* 41: 17–18.
Cubitt, Sean. 2016. *Finite Media: Environmental Implications of Digital Technologies*. Duke: Duke University Press.
Cubitt, Sean. 2019. "Ecocritique." *Media+Environment* 1 (1). https://doi.org/10.1525/001c.10784.
de Luca, Tiago, and Matilda Mroz. 2023. "Elemental World Cinema." *Studies in World Cinema* 3 (1): 1–18. https://doi.org/10.1163/26659891-03010001.
Dernikos, Bessie P. 2018. "'It's like You Don't Want to Read It Again': Exploring Affects, Trauma and 'Willful' Literacies." *Journal of Early Childhood Literacy*: 1–32. https://doi.org/10.1177/1468798418756187.
Deuze, Mark. 2012. *Media Life*. Cambridge: Polity Press.
Duncan, Pansy. 2023. "The Eco-material Posthuman in the Age of the Anthropocene." In *Screening the Posthuman*, Missy Molloy, Pansy Duncan, and Claire Henry, 244–275. Oxford and New York: Oxford University Press.
Edwards, Justin D., Rune Graulund, and Johan Höglund, eds. 2022. *Dark Scenes from Damaged Earth: The Gothic Anthropocene*. Minneapolis: University of Minnesota Press.
Escobar, Cristóbal, and Barbara Creed. 2024. "Becoming Nonhuman." Special issue "Film and the Nonhuman." *Senses of Cinema* 109. https://www.sensesofcinema.com/2024/film-and-the-nonhuman/becoming-nonhuman/.
Fay, Jennifer. 2018. *Inhospitable World: Cinema in the Time of the Anthropocene*. Oxford and New York: Oxford University Press.
Ferdinand, Malcom. 2019. *Decolonial Ecology: Thinking from the Caribbean World*. Cambridge: Polity Press.
Ghosh, Amitav. 2016. *The Great Derangement: Climate Change and the Unthinkable*. Chicago: Chicago University Press.
Haraway, Donna. 2016. *Staying with the Trouble: Making Kin in the Chthulucene*. Durham: Duke University Press.
Haraway, Donna, and Marta Segarra. 2019. *El món que necessitem / The World We Need*. Barcelona: Centre de Cultura Contemporània de Barcelona.
Ivakhiv, Adrian J. 2013. *Ecologies of the Moving Image: Cinema, Affect, Nature*. Waterloo, Ontario: Wilfrid Laurier University Press.
Kaplan, E. Ann. 2016. *Climate Trauma: Foreseeing the Future in Dystopian Film and Fiction*. New Brunswick, New Jersey: Rutgers University Press.
Kara, Selmin. 2016. "Anthropocenema: Cinema in the Age of Mass Extinctions." In *Post-Cinema. Theorizing 21st-Century Film*, edited by Shane Denson and Julia Leyda, 750–84. Falmer: REFRAME Books.
Kolbert, Elizabeth. 2014. *The Sixth Extinction: An Unnatural History*. New York: Picador.
Landreville, John. 2019. "Alternative Ethics for Ecocinema: *The Tree of Life* and Post-Continuity Aesthetics." *New Review of Film and Television Studies* 17 (1): 1–21. https://doi.org/10.1080/17400309.2019.1563365.
Leikam, Susanne, and Julia Leyda. 2017. "Cli-Fi and American Studies: An Introduction." *Amerikastudien* 62 (1): 109–120.
LeMenager, Stephanie. 2014. *Living Oil: Petroleum Culture in the American Century*. New York: Oxford University Press.
Leyda, Julia. 2023. *Anthroposcreens: Mediating the Climate Unconscious*. Cambridge: Cambridge University Press.

López, Antonio, Adrian Ivakhiv, Stephen Rust, Miriam Tola, Alenda Y. Chang, and Kiu-wai Chu, eds. 2024. *The Routledge Handbook of Ecocinema Studies*. New York: Routledge.
MacDonald, Scott. 2013. "The Ecocinema Experience." In *Ecocinema Theory and Practice*, edited by Stephen Rust et al., 17–42. New York: Routledge.
Marks, Laura. 2020. "Let's Deal with the Carbon Footprint of Streaming Media." *Afterimage* 47 (2): 46–52.
McKim, Kristi. 2015. *Cinema as Weather*. London: Routledge.
McMahon, Laura. 2014. "Beyond the Human Body: Claire Denis's Ecologies." *Alphaville* 7 (7): 94–111.
Molloy, Missy, Pansy Duncan, and Claire Henry. 2023. *Screening the Posthuman*. Oxford and New York: Oxford University Press.
Monani, Salma. 2024. *Indigenous Ecocinema: Decolonizing Media Environments*. Morgantown, WV: University of West Virginia, Salvaging the Anthropocene series.
Moore, Jason. 2015. *Anthropocene or Capitalocene?: Nature, History, and the Crisis of Capitalism*. Oakland: PM Press.
Morton, Timothy. 2009. *Ecology Without Nature: Rethinking Environmental Aesthetics*. Cambridge, Mass.: Harvard University Press.
Morton, Timothy. 2010. *The Ecological Thought*. Cambridge, Mass: Harvard University Press.
Morton, Timothy. 2024. "Rejecting the Anthropocene is a mistake." *IAI News*, April 9, 2024. https://iai.tv/articles/rejecting-the-anthropocene-is-a-mistake-auid-2806.
Narine, Anil. 2018. *Eco-Trauma Cinema*. London: Routledge.
Nixon, Rob. 2011. *Slow Violence and the Environmentalism of the Poor*. Cambridge, Mass: Harvard University Press.
Oppermann, Serpil. 2018. "The Scale of the Anthropocene: Material Ecocritical Reflections." *Mosaic* 51 (3): 1–17.
Paszkiewicz, Katarzyna. 2021. "Cinema and Environment: The Arts of Noticing in the Anthropocene." *Res Rhetorica* 8 (2): 2–21.
Paszkiewicz, Katarzyna, Andrea Ruthven, and Marta Segarra. 2021. "Posthuman kinship." *Dictionary of Gender in Translation*. https://worldgender.cnrs.fr/es/entradas/posthuman-kinship/.
Pesses, Michael W. 2021. *Ecomobilities: Driving the Anthropocene in Popular Cinema*. Washington: Lexington Books.
Pick, Anat. 2011. *Creaturely Poetics: Animality and Vulnerability in Literature and Film*. New York: Columbia University Press.
Rust, Stephen, Salma Monani, and Sean Cubitt, eds. 2023. *Ecocinema Theory and Practice 2*. New York: Routledge.
Seymour, Nicole. 2018. *Bad Environmentalism: Irony and Irreverence in the Ecological Age*. Minneapolis: University of Minnesota Press.
Simmons, Kali. 2019. "Reorientations; or, an Indigenous Feminist Reflection on the Anthropocene." *Journal of Cinema and Media Studies* 58 (2): 174–179.
Telles, Jason Paolo. 2024. *Indigenous Media and Popular Culture in the Philippines*. London: Palgrave.
Tidwell, Christy, and Carter Soles, eds. 2021. *Fear and Nature: Ecohorror Studies in the Anthropocene*. Philadelphia: Pennsylvania State University Press.
Todd, Zoe. 2016. "An Indigenous Feminist's Take On The Ontological Turn: 'Ontology' Is Just Another Word For Colonialism." *Journal of Historical Sociology* 29 (1): 4–22.
Troon, Simon R. 2024. *Cinematic Encounters with Disaster: Realisms for the Anthropocene*. London: Bloomsbury.
Uy, Jamie, and William Brown. 2020. "Cancelling the Apocalypse: *Pacific Rim* as Chthulucinema." *New Review of Film and Television Studies* 18 (4): 383–401. https://doi.org/10.1080/17400309.2020.1801311.

Watts, Vanessa. 2013. "Indigenous Place-Thought & Agency Amongst Humans and Non-humans (First Woman and Sky Woman Go on a European World Tour!)." *Decolonization: Indigeneity, Education & Society* 2 (1): 20–34.

Weik von Mossner, Alexa, ed. 2014. *Moving Environments: Affect, Emotion, Ecology, and Film*. Waterloo, Ontario, Canada: Wilfrid Laurier University Press.

Weik von Mossner, Alexa. 2017. *Affective Ecologies: Empathy, Emotion, and Environmental Narrative*. Columbus: The Ohio State University Press.

White, Daniel. 2019. *Film in the Anthropocene: Philosophy, Ecology, and Cybernetics*. London: Palgrave.

Whyte, Kyle. 2017. "Indigenous Climate Change Studies: Indigenizing Futures, Decolonizing the Anthropocene." *English Language Notes* 55 (1–2): 153–162.

Willoquet-Maricondi, Paula, ed. 2010. *Framing the World: Explorations in Ecocriticism and Film*. Charlottesville: University of Virginia Press.

Yusoff, Kathryn. 2018. *A Billion Black Anthropocenes or None*. Minneapolis, MN: University of Minnesota Press.

Zhong, Raymond. 2024. "Geologists Make It Official: We're Not in an 'Anthropocene' Epoch." *The New York Times*, March 20, 2024.

Part 1

Affect, Ecology, and Pedagogies of Worldly Reciprocity

2 A Film History of Utter Rebellion

Dewesternizing Film Studies for the Chthulucene

William Brown

This chapter takes its inspiration from a course on early cinema that I have been teaching over the past couple of years at the University of British Columbia, and it draws upon the attempt on my part as a course instructor to dewesternize/decolonize the course's curriculum. Having inherited the course, which is a second-year intro to pre-sound cinema, I chose, outrageously perhaps, to omit various key filmmakers and/or movements from the syllabus and instead to focus near-exclusively on 'minority' and non-western films and filmmakers. While mention and some explanation is of course made of D.W. Griffith, Soviet montage, German Expressionism, and male-authored French experimental cinema of the 1920s, these filmmakers and movements are nonetheless bypassed, meaning that the course overlooks numerous canonical texts that would otherwise be standard fare in an 'intro' class such as this. Rather than tell the canonical story of film history (as per many an 'intro'), the twin aims of the course are to engage both with how/whether cinema has always been figuring the end of the world (and thus is always already expressing anxiety around the meaning of the Anthropocene), and to demonstrate that the widespread availability of early cinema on YouTube and other online archives allows us to reorder the canon and/or to rewrite film history from different perspectives. In this way, the idea is for students to investigate whether a different film history is possible, one that might thus point to a different cinema, which in turn might help us to 'think' a different world (involving, perhaps, a non-destructive approach to our planet, or a world 'beyond' the Anthropocene and in/of what Donna J. Haraway might term the 'chthulucene'; see Haraway 2016).

Decolonizing the Curriculum

It potentially goes without saying that the course comes up immediately against both historical and ongoing imbalances of power across the globe. Which places have been able to preserve and/or restore the films that were made and/or exhibited there, and in what condition are the films that persist, including whether they have been translated? Indeed, which films have been digitized at all, let alone also uploaded to a website to which we have access?

DOI: 10.4324/9781003468028-3

Furthermore, of those films that are available, to what extent might the one among several YouTube versions that I used in class of, say, *Aelita, the Queen of Mars* (Yakov Protazanov, USSR, 1924), be understood as 'official' or 'correct'—especially when the other versions have different running times and/or soundtracks? That is, part of the course involves embracing and working with, rather than against, the fragmentary and possibly 'incomplete' (or even reworked) nature of some of the films that we see, in the process challenging the very mentality and/or desire for 'completion' and/or 'officialdom' that is endemic of the west, and which cannot but be fraught with political biases and issues. That is, the very desire for complete, official and/or 'static' versions of films is itself redolent of a western outlook—even as it runs contrary to the very medium in which these films are made, in that if film is anything, it is not static, as film historians like Paolo Cherchi Usai make clear when they suggest that there is no real 'original' (and thus authoritative) version of a film (2010, 160). By this twin rationale, the course, which Katherine Groo (2019) might consider an example of 'bad film history,' aims to ask if one of the potentials of digital technology is that it liberates us from the strictures of completism and canon—even if it is sadly hard to find (and thus to include in the syllabus) any sub-Saharan African-authored silent cinema, for example, perhaps because it has never existed (which begs an important question, also raised by Neepa Majumdar, regarding what constitutes 'early cinema' and to whom; see Majumdar 2013; and Kumar 2013, 153).

The course is not without western texts, which do indeed constitute a large proportion of the films included. Nonetheless, even the western texts are intended to be unconventional choices for an intro course, as, after looking at various actualities, trick films and (proto-)narrative shorts, our first encounter with feature-length cinema is *Verdens Undergang/The End of the World* (August Blom, Denmark, 1916), which shows that planetary destruction has been on western (and perhaps other) minds for at least a century. We then look at *Cabiria* (Giovanni Pastrone, Italy, 1914) to remind ourselves that the narrative feature does not originate in the United States (while also being a film without an 'original' version; see Cherchi Usai and Sopocy 1988). The afore-mentioned *Aelita …* provides us next with a Soviet perspective on the extra-planetary, while *White Fawn's Devotion* (James Young Deer, United States, 1910), *Daughter of Dawn* (Norbert Myles, United States, 1920) and *Ramona* (Edwin Carewe, United States, 1928) all offer up Indigenous North American perspectives on the relationship between man and the natural world—with the latter including an extraordinary scene in which the title character (Dolores del Río) receives word of her Native ancestry not with the horror expected by her adoptive mother, Señora Moreno (Vera Lewis), but with total joy, since it means that she can marry her love, fellow Native and farm hand, Alessandro (Warner Baxter).

We then look at *The Navigator* (Buster Keaton and Donald Crisp, United States, 1924), but not so much in order to 'cover' early silent comedians as to engage with the white supremacy at work in this more 'Hollywood' fare, including how the film relates to what I have elsewhere termed, *chapeau à* Haraway, the transition from the white Anthropocene to the Black Chthulucene (see Brown 2023). We also look at Black and white women's filmmaking in North America through, respectively, *Within Our Gates* (Oscar Micheaux, United States, 1920) and *The Ocean Waif* (Alice Guy Blaché, United States, 1916), before then thinking about early cinema through a white feminist-apocalyptic lens via *Häxan: Witchcraft Through the Ages* (Benjamin Christensen, Sweden/Denmark, 1922) and *La Souriante Madame Beudet/The Smiling Madame Beudet* (Germaine Dulac, France, 1923). We consider the remaining fragments of *The Story of the Kelly Gang* (Charles Tait, Australia, 1906) to engage most directly with 'incompleteness,' especially as its existence as 'ruin' reminds us that our world has only finite resources (see Cubitt 2017, 1–12), while also demonstrating that the 'western' as a genre is perhaps not so 'western' (as in, occidental) as we might presume.

Finally, we consider *Barsoum cherche un emploi/Barsoum Looking for a Job* (Mohamed Bayoumi, Egypt, 1923), *Kurutta ippêji/A Page of Madness* (Kinugasa Teinosuke, Japan, 1926), *Kaliya Mardan* (Dadasaheb Phalke, India, 1919), *Shiraz* (Franz Osten, India/United Kingdom/Germany, 1928), *Hongxia/Red Heroine* (Wen Yiming, China, 1929), *Nobleza gaucha* (Eduardo Martinez de la Pera, Ernesto Gunche and Humberto Cairo, Argentina, 1915), and *El tren fantasma* (Gabriel García Moreno, Mexico, 1926) in order to introduce early cinema from a range of places, thereby demonstrating the medium's global reach and expression. The intention is not to clump all of these latter, non-western films together, as if they were the same. Indeed, the radicality of Kinugasa's film is barely matched in *El tren fantasma*, for example, which would rather be a normative celebration of Mexican nationalism (although such nationalism can take on a relatively subversive quality when pitted against the cinematic imperialism of Mexico's northern neighbor, something mirrored in our consideration of *Kaliya Mardan* below). But, the very diversity of non-western films screened in this second half of the course attempts to break down Euro-American hegemony and canonical repetition from without, much as the first half aims to do so from within. Disparate as they are, then, there is nonetheless an opportunity to seek commonalities among these films, perhaps especially in terms of their relationship with the western modernity that gave birth to cinema, and which thus indelibly shapes these films for better and/or for worse.

Uncontained Digital Archives

As mentioned, all of the films are available online, and we project them directly from the internet in class, deliberately eschewing 'better' versions of the films as stored on DVD or BluRay (although we have on occasion included a 16mm projection of Buster Keaton and Edward Cline's *The Playhouse*, United States, 1921, to get a sense of watching material film stock on a big screen—with the rationale behind this title simply being that it is one of only a few in UBC's 16mm collection).[1] More to the point, though, is that the endeavor to dewesternize the syllabus is also an endeavor to decolonize the western perspective, leading hopefully to a sense that western filmmaking presents to us not just stories but also a worldview (and/or a way of telling those stories) that are not timeless and universal, but rather historically contingent. That is, if "the writing of History—a combination of words, objects, and events—is a *poetic*, creative act like any other discursive practice" (Groo 2012), then the course seeks 'poetically' to write new or alternative histories of cinema. And we do this because if the world is coming to an end, then not only does it seem that cinema has long since known this—as have Native Americans and those whose worlds were destroyed by the Middle Passage and other forms of (especially gendered) persecution and colonialism—but it also is to suggest that decolonization is perhaps our best chance of doing something about it, or at the very least learning to accept transience and death (rather than to deny them, which is white western modernity's usual practice—a practice inscribed into the medium of film as one in a line of technologies used to extend life beyond the physical bounds of the body, even as film decays, and even as film might have a life of its own above and beyond the lives of those it documents).

Given both its role in preserving time (Bazin 2005, 9–16) and in globalizing American ideas and products (as per Frank J. Marion's oft-cited declaration in 1918 that "[t]rade follows the film;" see Thompson 1986, 122), it might seem that cinema has been a tool for westernization and the instigation of the destructive Anthropocene. Indeed, the role of cinema in the 'Americanization' of the world ('globalization') seems undeniable, with that same 'Americanization' bringing us to the brink (if not beyond the tipping point) of environmental cataclysm. This 'Americanization' is not limited to a set of practices, including but by no means restricted to making films; rather, it also comprises a set of attitudes, or a worldview, or a philosophy, as well as an ensuing relationship (or lack thereof) with the world and the lifeforms that are entangled with it (including other humans). As we shall see, this philosophy or (lack of) relationship is one defined by a sense of separation (i.e. a relationship of no relation), which in turn enables exploitation (I am not connected to the planet or to these people, and so I can treat them as I wish). Linked to this sense of separation is consequently a predilection for 'closure' (which, in opposition to openness, marks the point of separation from a world or other that otherwise threatens the westerner), with western stories generally seeking closure/resolution (or achieving stasis), which in turn drives the medium in its dominant iterations toward being a narrative form.

In contrast to closure, stasis and narrative, the syllabus aims to highlight stories that are significantly more open, unstable and which shift away from narrative—even as they do not dispense with the latter altogether. Furthermore, beyond the relationship that many of these films have with openness (or their open relationship with relationship itself, whereby they are open to openness), we find that the digital archive, with its multiple unofficial versions, its fragmentary nature and its incompleteness, also helps to convey the lack of closure that is inherent to a medium that only became dominated by narrative and an ethos of closure some ten years or so after its invention (see Gunning 2006 [1986]). The digital archive can thus be understood as what Rashmi Devi Sawhney would term an "uncontained" archive, in that it "negates the totalitarianism of the colonial archive … [and] is incomplete and always in the process of being built and rebuilt." This "fluidity" in turn "allows us to think of historical narratives as projections into present times, rather than as entities belonging to a distant and cut-off past" (2023, 216). That is, the digital archive, especially when working in conjunction with a dewesternized syllabus, allows us ('poetically,' *à la* Groo) to wrestle a new understanding of cinema from historically dominant filmmaking practices, while also helping us to clarify an openness that has always been there in cinema, but which has been occulted as a result of western hegemony in the world, in dominant film production, and in strands of film studies. In this way, the course allows us to construct alternative histories from which we can learn as we live in a present defined by an uncertain future. In short, we do not need to use film history as a means to gain control or a(n intro-level) mastery over cinema; rather, we can unleash and learn to live with the 'trouble' (Haraway 2016) of cinema's chaotic tendencies, as intensified by the turmoil retroactively engendered by a digital technology that might otherwise promise only more control.

Drawing from experimental Japanese author Yokomitsu Riichi's desire to fight "a war of utter rebellion against the Japanese language," an attitude that influenced Kinugasa's afore-mentioned film *Page of Madness*, which in turn waged a "war of utter rebellion against film language" (see Peterson 1989, 38 and 51), this chapter wishes therefore to describe and to contribute to a 'film history of utter rebellion,' or a 'war of utter rebellion against film history (as it is typically recounted).' The aim is not to diminish the 'rebellious' qualities of numerous films *within* the canon of Euro-American cinema (no one is trying to deny that *Nosferatu: eine Symphonie des Grauens*, F.W. Murnau, Germany, 1922, or *Un chien andalou/An Andalusian Dog*, Luis Buñuel and Salvador Dalí, France, 1929, are 'great' films—even if the course does not provide the constant reaffirmation of such 'greatness' that some seem to feel necessary). But it is to seek to open up that same Eurocentric canon (to share the 'greatness' around a bit more?), as well as to open up the attempted closure that is canon-formation itself (only certain films qualify as 'great,' typically as a result of a consensus believed to be objective rather than manufactured). And so, even if we are 'only' talking about a course on early cinema, which might rank low on most people's list of urgent things to do in the face of the end of the world, the aim here is to contribute to the decolonial effort—and the rest of this chapter aims to clarify by way of comparison the afore-mentioned 'philosophical' differences between texts eastern and western, even if it can only do so in relatively broad brushstrokes (as befits the 'intro' level).

Kill or Kin?

Toward the end of *The Navigator*, Rollo Treadway (Keaton) dons a diving suit and heads underwater in order to try to liberate the film's titular ship, which has run aground after drifting pilotlessly, and with only two passengers, for an unspecified period of time. Among other things, Rollo is attacked underwater by an octopus—just as Pacific islanders, played in the film by Black actors, abduct his love interest and fellow passenger Betsy (Kathryn McGuire) from the *Navigator*. Rollo then stabs the octopus to death behind a rock, from which emerges a plume of ink and blood, before he surfaces, still in his diving gear, on to the shore of the unnamed island (Rollo was hoping to make it to Honolulu). His diving costume frightens the islanders, allowing Rollo to rescue Betsy and return to the ship. However, the Natives give chase, and Rollo attempts to fight them off, killing one of them with a mini-cannon that nearly fires directly at Rollo himself. Overrun by semi-naked Natives, Rollo and Betsy abandon ship and attempt to escape on an Indigenous canoe, which unfortunately sinks. Rollo and Betsy re-emerge from the water, however, on the foreplane of an American submarine, which rescues them, thereby allowing Rollo, who up until now had been rebuffed by Betsy in his requests to marry her, to take Betsy's hand and thus complete his heteronormative destiny as a white American man.

There is much to analyze in this sequence, as I have elsewhere demonstrated (Brown 2023). However, I invoke it here so as to compare and contrast it with the final moments of D.G. Phalke's *Kaliya Mardan*, sometimes referred to as *The Childhood of Krishna*, and which is an early example of what Rachel Dwyer (2006) identifies as the "mythological" genre—a specialty of its director, who is widely credited with being the first maker of fiction feature films in India. The film, or at least the 4,441 feet of it that remain (Dwyer 2006, 24; this equates to 48 minutes and 47 seconds of YouTube screen time), offers up episodes from the childhood of Krishna, and as such is in principle a *pauranic*, in that "Krishna's *lila* or life is told in the *Bhagavata Purana*" (Dwyer 2006, 15), even though many mythologicals draw more closely on India's "two great epics, the *Mahabharata* and the *Ramayana*" (16). The film's final sequence, described by Sudesh Chabria as "the most important reference point for any study of Phalke and the mythological genre today" (1993, 110), depicts how the child Krishna (Mandakini Phalke) defeats Kaliya, a Nāga (half-human, half-serpent) who according to legend comes to live in the Yamunā River in Vrindavan following his expulsion from the island of Ramaṇaka by Hindu deity Garuḍa, a renowned snake hunter. Indeed, Kaliya chooses Vrindavan as his home because a yogi, Saubhari, curses Garuḍa, who will die if he enters the city—meaning that Kaliya is in principle safe there. However, as Kaliya poisons the river and thus threatens the lives of the local population, so does Krishna decide to defeat the creature in order to prevent it from causing further harm. Now, while the river functions as a backdrop for various of the sequences of *Kaliya Mardan* that remain, none of this backstory is provided within the film. This is because Phalke's viewers would already have been very familiar with these events—a fact that leads Sean Cubitt to suggest that "[i]n Phalke ... one has the sense that the narrative, however fragmented by the vagaries of preservation and transmission, occupies a sovereign position, but as metonymic, almost holographic shards of the Hindu epic tradition" (1999, 124). That is, while there is ostensibly narrative in Phalke's film (see also Dwyer 2006, 22), it really stands in for a story that is already known, and which thus translates *cinematically* into a fragmented (or episodic) style.

We shall look at the aesthetics of *Kaliya Mardan* in more detail shortly, but for the time being we should recount how the film picks up the story with Krishna climbing a tree over the river, before diving in to do battle with the coils of the snake underwater. A group of locals look on from the bank, but given Krishna's prolonged disappearance, they fear that the child-god has died—and go to report as much (without intertitles) to Krishna's foster mother, Yashoda. News spreads of Krishna's death, and as people gather to mourn him, so does the film intercut with Krishna's struggles below the surface, with the child eventually arising from the waters atop the hood of the snake just as mourners gather on the river's banks. However, even though *Kaliya Mardan* is translated by Sawhney as *Death of Kaliya* (2023, 224), the child does not kill the snake, but rather spares (or, as most translations have it, "defeats") him as a result of the pleas from Kaliya's two wives, who also emerge from the water, superimposed like angelic figures onto the image of the snake and boy.

Now, even though "[s]nakes are not evil figures in Indian mythology but are auspicious and must be worshipped, especially on festivals such as Naag Panchami [a day in July/August for worshipping snakes]" (Dwyer 2006, 48), Kaliya nonetheless symbolizes "the traditional concept of evil in Hindu mythology"—as well as British rule, as we shall see (Chabria 1993, 114). Krishna still does not kill Kaliya, however, preferring instead, in the language of Donna J. Haraway, to make "kin" with the creature—or at least this is the reading that I wish to offer here by way of contrast to *The Navigator*. For, in the latter film Rollo kills the octopus and then one of the Natives in order to affirm his familial role in the normative white couple. What is more, *The Navigator* cuts from the octopus grabbing Rollo underwater to the Natives as they board the ship and grab Betsy, meaning that the film suggests via its editing an equivalence between the octopus and Blackness/Indigeneity. In other words, *The Navigator* establishes via its editing a narrative in which white normativity is built upon the simultaneous exclusion of the animal and the non-white other. Indeed, as we have extensively argued elsewhere, the octopus is understood popularly as an 'alien intelligence,' while also figuring prominently, along with other tentacular creatures, in the Cthulhu mythos of H.P. Lovecraft, which in turn functions, at least in the context of the United States, as an elision of the cephalopod with Blackness and other non-white identities during the Jim Crow era (see Brown and Fleming 2020; Brown 2023;

Brown and Fleming, forthcoming). That is, the octopus in *The Navigator*, which was made during the same period in which Lovecraft emerged as a recognized writer of 'weird tales,' gives expression to an anti-Black and anti-Indigenous worldview, and which requires the death of the octopus and the Black/Indigenous other for the purposes of the construction of the white heterosexual couple. Or, if you will, the realization of American Manifest Destiny does not have murder as an unfortunate and collateral side-effect; it is, rather, its very basis.

In contrast to *The Navigator*, the aquatic encounter with a cthulhoid/tentacular other in *Kaliya Mardan* does not (need to) result in murder (while Kaliya only has one head in Phalke's film, its 'cthulhoid' credentials do increase when we consider that in popular mythology it is a many-headed beast), suggesting a worldview that is distinct from the western one of separation and exploitation, and which is instead relational. Now, while *The Navigator* in its editing elides the animal with Blackness, *Kaliya Mardan* makes no equivalent association, even as the snake is associated with Britishness, as per Chabria's afore-mentioned reading. While we shall work with this interpretation below, the point simply to make is that even if the snake does represent (British) 'evil,' this is still not cause to kill it—and does not need to be, because 'Indian' identity, as opposed to 'white' identity, does not rely on murder in order to exist.

Anti-Colonial Phalke?

The goal here is not to romanticize or 'orientalistically' fetishize this work of 'Indian' cinema (the scare quotes will be explained) as an exotic other, overly burdening it as emblematic of an Indian society that never involves violence, which is in harmony with nature, and which thus holds the answers to all of the western world's problems. Indeed, we would do well to point out here that, while Phalke is credited as being the father of Indian cinema, there is evidence to suggest that he was not in fact the first local maker of films in India (examples from before Phalke's first feature, *Raja Harishchandra* from 1913, might include Hiralal Sen's short, *Ali Baba*, from c1903, and the feature-length film, *Pundalik*, by Dadasaheb Torne, from 1912; see Mukherjee 2020, 320). Furthermore, while the mythological was a key genre of the 1910s and onwards, mainly thanks to Phalke, whose *Raja Harishchandra*, *Lanka dahan/Lanka aflame* (India, 1917) and *Shree Krishna Janma/The Birth of Shri Krishna* (India, 1918) constitute key examples (with the remaining fragments of the latter also featuring Krishna atop Kaliya in its opening moments), there is an overemphasis on the genre's presence in Indian cinemas in the 1910s and 1920s (see Thomas 2014, 2; Mukherjee 2018, 213–214). Additionally, while Phalke's films were commercially successful, the audiences for mythologicals were mainly from the lower classes (Dwyer 2006, 56), and, as far as Phalke specifically is concerned, not everyone liked his work, as Dwyer makes clear when she cites Satish Bahadur, who argues that

> [f]or dramatics, Phalke drew upon the crude elements of the 'Company Natak' [a Pune-based theatre group] not the glories of classical Sanskrit drama or the vital forms of the folk theatre. For story material, Phalke did not delve deep into the spiritual meaning underlying the Hindu epics. Rather he used their most obvious ritualistic and superficial level, viz. the magical, the miraculous and the spectacular in the exploits of the Hindu Gods and goddesses.
>
> (Dwyer 2006, 55; see also Bahadur 1976, 91–92)

Finally, while Phalke's films might have been technically innovative, such that they both were seen outside of India and merit comparison with Georges Méliès, they have also been read as working "in a theological time in which the past subsists in the present as the gateway for the future … as a function of their attempt to legitimate and administer a specific post-colonial future in the name of an educated, nationalist, Hindu, and male elite" (Cubitt 1999, 126).

In other words, Phalke was not the first; his chief genre was not the only; his works were not universally popular; and his works arguably demonstrate a middle-class, patriarchal and specifically Hindu (as opposed to, say, Muslim) ideology, even as they were mainly popular among a lower class demographic domestically, while also drawing some foreign audiences (who may have been attracted to them because they pandered to the western, orientalist appetite for Indian magic and miracles), which in turn highlights the limits of his credentials as essentially an 'Indian' filmmaker (hence the scare quotes).[2]

This being said, Phalke considered his practice to be *Swadeshi*, which was a movement toward Indian independence, nationalism and, in particular, self-sufficiency, which, in the filmmaking context, involved using a foreign technology, the cinematograph, to tell Indian stories. For Ashish Rajadhyaksha, it is Phalke's *Swadeshi* approach that allows him to resolve all of the above contradictions (1987, 64), while M. Sadiqur Rahman sees Phalke as specifically an "anti-imperial" filmmaker (2019, 532) who is working in a context that is "already antipathetic to colonialism" (Cubitt 1999, 123). Read in this light, the sequence from *Kaliya Mardan* "represents the total adaptation and indigenisation of the imported medium of cinema to Indian filmic and cultural practice" (Chabria 1993, 114). That is, Phalke creates an 'Indian' cinema that is defined through a colonial history, and thus which is in some senses in opposition to the colonizer—most obviously the British, but also the technology of cinema itself, which could be described as a Franco-American invention (the parallel claims of the Lumière brothers and Thomas Edison to be its inventors), even though Phalke bought his equipment from British cinema pioneer Cecil Hepworth on a trip to the UK in 1912 (Dwyer 2006, 22; Rahman 2019, 524). However, this 'opposition' is not total, not least because Phalke still *makes films*; as Rajadhayaksha says, Phalke "wanted to prove to all (i.e. the English) that an Indian working under primitive conditions could make films too" (Rajadhyaksha 1987, 48). As such, there is what Rosie Thomas calls "a broader field of coeval cosmopolitan filmic influences and impulses" at work in Hindu mythological epics, and by extension Phalke's work (2014, 2; also cited in Mukherjee 2020, 320).

As Valentina Vitali says, Hindu mythology is not so much "in tension with the notion of Indian modernisation ... as a factor instrumental to its realisation" (2008, 8; also cited in Mukherjee 2020, 213). In other words, the construction of 'India' does not come about as a result of the death, or murder, of the colonizer—even as modernity only comes about through the colonizer's murder of the racialized other. Instead, 'India' is shown, or proves itself, to be equally as 'cinematic' (i.e. as capable of cinema, or as capable of being modern) as the colonizing forces (the United Kingdom, but also France, the United States and other western powers), even if it speaks of and to that modernity from its 'flipside.' The idea, then, would not be to appeal to India as if it were a 'premodern' space from which we (westerners) might learn how to live *as if modernity never took place*. Phalke functions, rather, as a filmmaker who was always, and could only be, reconciling modernity with his own 'subaltern' position within it (Spivak 1994), if not his exclusion therefrom. It is because of his situation in precisely this space—outside of modernity but not separate from it—that Phalke can, in Cubitt's comparison between the Indian filmmaker and Méliès, upend "the presumed chronological relation between European master [Méliès] and Indian disciple ... so that the colony precedes the metropolis" (1999, 116). Similarly, it is precisely because Phalke cannot but make films otherwise (in a fashion other to that of the west) that he can help us to imagine an other cinema, and which otherness is expressed precisely in its relationality with otherness (Krishna with the snake), as opposed to its opposition thereto (Rollo kills the octopus and the Native).

If we cannot undo modernity, then from within it we (westerners) must find a way to live otherwise, since the ongoing murder and exploitation of the (racialized) other—the Anthropocene—is leading to our mutual and collective destruction. If Cubitt says that, "[a]s Other, Phalke speaks from the place of an internalised outside, an exotic space which *only appears* to negate the colonial-universal imagination" (1999, 127; emphasis added), then we might slightly disagree; for while it may be that Phalke does not *negate* the colonial-universal imagination (such that it no longer exists), his liminal situation can, I argue, give us fodder for rumination—to unsettle typical western thinking and thus to decolonize. Or at least, as a western educator invested in decolonization, I believe that Phalke's *Kaliya Mardan* can, in its relationality, provide us (me) with an opportunity to learn how to be (more?) relational—in the belief indeed that the Indian (as well as the other non-western filmmakers that we consider) is our master, and we the disciples (especially when it comes to learning to 'stay with the trouble').

The Aesthetics of Relationality

In his comparison of Phalke with the transnational (typically German-British-Indian) co-productions of Himanshu Rai, which include the film *Shiraz* mentioned above as part of the early cinema syllabus, Rahman suggests that while both work within the same anti-imperial context, it is Franz Osten's film that in particular is created for the tourist gaze, while Phalke's is not (2019, 538). Our concern is not whether this makes Phalke more 'Indian' (and here perhaps we agree with Dwyer in that "nationalism has been overdetermined as part of the viewing of Indian cinema;" see Dwyer 2006, 59). However, it is to suggest that Phalke is less concerned with 'pandering to the orientalist west' than a skeptic might be tempted to suggest.

Meanwhile, Debashree Mukherjee observes that there is

> an enduring perception of early Indian audiences as aesthetically naïve image-worshippers. Instead, it serves us well to consider other experiments, such as Phalke's own short *How Films Are Made* (1917), [in order] to locate a self-reflexive impulse in the earliest years of cinema in India. Placed alongside the opening sequence of *Kaliya Mardan*, where Phalke's daughter Mandakini reveals her acting repertoire and her costume-dependent transformation into the boy Krishna, Phalke positions cinema as an art-industrial form capable of visual illusionism.
> (2020, 212)

Not only does Mukherjee's argument push back against any suggestion that the 'lower class' audiences of Phalke's films might similarly be 'naïve' (that is, that not just Indian, but especially *poor* and thus uneducated Indian audiences are 'naïve'), but it also suggests a self-consciousness about Phalke's work—as well as, perhaps, an 'Indianness' in that the sequence stages "the Indian aesthetic theory of *navarasas* or the nine emotional aesthetic moods inscribed in Indian classical arts" (Chabria 1993, 115). And so, if Rahman criticizes the long shots of, say, *Raja Harishchandra* as not allowing us to get closer access to the emotional state of the characters (while *Shiraz* does feature such shots; see Rahman 2019, 527), we might less consider this to be a 'weakness' of Phalke's work than it is a deliberate ploy. And why might Phalke do this? Because longer shots allow us precisely to see the characters *in relation to*, as opposed to isolated from, their surroundings—with Phalke's decision to shoot on location in his hometown of Nasik rather than in a bigger city also being an intentional strategy in order to differentiate (and make *Swadeshi*) his work (see Dwyer 2006, 24–25).

With reference to the opening sequence of *Kaliya Mardan*, Rajadhyaksha also writes that

> [i]n Phalke there is almost no definition of time; the contiguities are effected in the different states of seeing as they come together, as they are activated by the static icon. The story, if there is one, is a continuous back-and-forth interaction between the viewers and the object viewed; we are shown the imaginary universe condensed into the object, our seeing is reciprocated. In *Kaliya Mardan* ... there is a long initial sequence where the actress Mandakini Phalke, the director's daughter,

> demonstrates her histrionic ability on screen before starting her role as Krishna. She, like the "mountains, rivers, oceans, houses..." is *real*: her reality is used to establish the illusion that Krishna must be real too. The true reality in all this is of course the audience and their desires.
> (1987, 71)

In other words, Phalke's episodic structure (moving away from narrative closure and/or a total explication of events) is matched by his use of long shots that function as, or bear similarities to, *tableaux* (no doubt a style derived from Phalke's training in the arts, as well as his apprenticeship under 'kitsch' painter Raja Ravi Varma; for more, see Rajadhyaksha 1987, 61–64). The 'frontality' of much of Phalke's cinema also 'implicates' the audience, as does the repeated gaze of Phalke's actors at the camera (see Chabria 1993, 110). Not only does Phalke "position cinema as an art-industrial form capable of visual illusionism," as Mukherjee (2020, 212) has it, then, but he also stages magic, or miracles, such as Krishna riding Kaliya to the surface of the Yamunā, as if they were 'everyday' events, thereby inducing a converse effect, which is to see the everyday as magic (or to see that magic and the everyday cannot be disentangled). This is equally a 'relational' technique, in that the viewer sees reality and myth as entangled or intertwined, such that myth (and/as cinema) can function as a means of helping us to understand the constructed (as well as 'magical') nature of reality, which in turn means that we can think and perhaps realize a new reality, one that is less destructive of our world, which in our present moment might be much required.

What is more, the audience is 'relationally' entangled into this process, in that the 'back-and-forth' described above by Rajadhyaksha is "a continuing projective identification into the site of crisis. It means a wish-fulfilment of appropriating the happy ending realized by external intervention (God/us) that finally places *us* back at the frontal helm" (1987, 76). Dwyer makes a similar case when she says that Phalke's *tableau* aesthetic "invites *darshan*," or "seeing," whereby in "the exchange of looks, the eye is an organ of tactility," and which expresses "a desire to fuse the image and the beholder" (Dwyer 2006, 19). That is, we are deliberately entangled with Phalke's film, which itself depicts entanglements of human and animal, myth and reality, and so on. In short, relationality. If Cubitt argues that "[s]uch magic [in Phalke] poses itself as the obverse of European enlightenment reason, but retains from it the patriarchal right of rule" (1999, 127), then we might again contradict him slightly to say not only that Phalke goes some way to mitigating this by casting a girl, his daughter Mandakini, as Krishna (as well as being a pioneer of working with women in general), but that it is also fitting for him to do so.

Conclusion

It is not that we should expect some sort of 'moral perfection' in Phalke or any filmmaker (we are not trying to negate all and any lingering patriarchal and/or exclusively nationalist elements in his work). However, if "Phalke's films have rarely been seen in Europe and North America … [if] he scarcely figures in non-Indian histories, and [if] the films themselves exist only in fragments now" (Cubitt 1999, 128), then perhaps we can nonetheless give to Phalke and his work something of a platform, and allow the anti-colonial work that he does to help us to decolonize our (white, western) selves. Phalke's work is not just anti-colonial in that he depicts specifically 'Indian' myths; it is also an early example of a 'relational' and 'open' cinematic style that can be understood not just as a 'poor imitation of' or 'inferior to' the dominant western style, but rather as a mode of ('subaltern') cinematic 'speaking' to which we would do well to listen, since it perhaps teaches us more (and more pressing things) than many a canonical western text can or will do. More particularly, it is not this way by accident, but deliberately and a result of various choices made by Phalke during the historical moment in which he found himself. The fragmented existence of *Kaliya Mardan* in the 'uncontained' digital archive of YouTube paradoxically enhances the power of the film, especially when both the condition and the content of *Kaliya Mardan* are read alongside the over-represented and over-preserved texts of the western canon.

If Rollo and Betsy venture outside only to be confronted by threatening Black/Indigenous others, as well as a cthulhoid octopus, then their reward/salvation at the end is to be reunited and thus to create the heteronormative home. Krishna, meanwhile, plays outside with his friends, and/or spends a lot of time mischievously breaking into people's homes. For example, Krishna breaks into a woman's home and steals her butter after she has soaked Krishna's friends with water, while also sneaking into a sleeping couple's room and tying the man's beard to his wife's hair. Not only does Krishna here literally entangle the couple, but he also seeks consistently to undermine the perceived security of the 'closed' space (the domestic hearth), and to remind those with whom he interacts of the outside (a trope that in the west becomes reworked as the horror staple of the home invasion movie). Maybe a film history of utter rebellion can also teach us to be mindful of (rather than threatened by) the outside, and the instability of things, not in such a way that we seek to control them to ever-greater degrees, but rather in such a way that we can, like Krishna, playfully work with them. Such a 'relational' ethics might be precisely what a 'non-canonical' or dewesternized film history can teach us, decolonizing our thought and helping us to navigate from the Eurocentric Anthropocene era and toward a Chthulucene in which we make kin with, rather than killing, others.

Notes

1 The course also typically involves at an early stage a visit to Rodney Graham's Millennial Time Machine (Canada, 2003), a camera obscura on the grounds of UBC. Focused on a sequoia tree that stands next to the university's Indian Residential School History and Dialogue Centre, the work provides an opportunity for students to begin reflecting on the relationship between imaging technologies and settler colonialism within a Canadian space.

2 Sawhney brings to our attention the work of Fatma Begum, a woman of likely Ethiopian descent who started making films in India shortly after Phalke. While none of her films are extant, Sawhney's emphasis on Begum's African heritage helps to complicate our understanding of Indian cinema as somehow being 'purely' 'Indian.' The aim of this chapter, and of my early cinema course, is not to make any claims to or for purity, but to encourage us to work positively with impurity.

Works Cited

Bahadur, Satish. 1976. "The context of Indian film culture." *Film Miscellany* (December): 90–107.

Bazin, André. 2005. *What is Cinema? Volume 1*. Translated by Hugh Gray. Berkeley: University of California Press.

Brown, William. 2023. *Navigating from the White Anthropocene to the Black Chthulucene*. Winchester: Zer0 Books.

Brown, William, and David H. Fleming. 2020. *The Squid Cinema from Hell:* Kinoteuthis Infernalis *and the Emergence of Chthulumedia*. Edinburgh: Edinburgh University Press.

Brown, William, and David H. Fleming. Forthcoming. *Infinite Ontologies of the Chthulustream:* ∞O. Edinburgh: Edinburgh University Press.

Chabria, Sudesh. 1993. "D.G. Phalke and the Méliès Tradition in Early Indian Cinema." In *Georges Méliès: Magier der Filmkunst*, edited by Frank Kessler, Sabine Lenk and Martin Loiperdinger, 103–115. Basel: Stroemfeld/Roter Stern.

Cherchi Usai, Paolo. 2010. *Silent Cinema: An Introduction*. London: British Film Institute.

Cherchi Usai, Paolo, and Martin Sopocy. 1988. "'Cabiria,' an Incomplete Masterpiece: The Quest for the Original 1914 Version." *Film History* 2 (2): 155–165.

Cubitt, Sean. 1999. "Phalke, Méliès, and Special Effects Today." *Wide Angle* 21 (1): 115–130.

Cubitt, Sean. 2017. *Finite Media: Environmental Implications of Digital Technologies*. Durham, NC: Duke University Press.

Dwyer, Rachel. 2006. *Filming the Gods: Religion and Indian Cinema*. London: Routledge.

Groo, Katherine. 2012. "Cut, paste, glitch and stutter: remixing film history." *Frames Cinema Journal* 1 (1). https://framescinemajournal.com/article/cut-paste-glitch-and-stutter/.

Groo, Katherine. 2019. *Bad Film Histories: Ethnography and the Early Archive*. Minneapolis: University of Minnesota Press.

Gunning, Tom. 2006. "The Cinema of Attraction[s]: Early Film, Its Spectator and the Avant-Garde." In *The Cinema of Attractions Reloaded*, edited by Wanda Strauven, 381–388. Amsterdam: Amsterdam University Press.

Haraway, Donna J. 2016. *Staying with the Trouble: Making Kin in the Chthulucene*. Durham, NC: Duke University Press.

Kumar, Ramesh. 2013. "On Scavenging and Salvaging: NFAI and Early Indian Cinema." *Framework: The Journal of Cinema and Media* 54 (2): 153–157.

Majumdar, Neepa. 2013. "What is 'Early' Cinema?" *Framework: The Journal of Cinema and Media* 54 (2): 136–139.

Mukherjee, Debashree. 2020. *Bombay Hustle: Making Movies in a Colonial City*. New York: Columbia University Press.

Mukherjee, Madhuja. 2018. "The Forking Paths of Indian Cinema: Revisiting Hindi films through their regional networks." In *The Routledge Companion to World Cinema*, edited by Rob Stone, Paul Cooke, Stephanie Dennison, and Alex Marlow-Mann, 213–224. London: Routledge.

Peterson, James. 1989. "A War of Utter Rebellion: Kinugasa's 'Page of Madness' and the Japanese Avant-Garde of the 1920s." *Cinema Journal* 29 (1): 36–53.
Rahman, M. Sadiqur. 2019. "Early 'Glocalization' in Indian Cinema: An Analysis of Films of Dada Saheb Phalke and Himanshu Rai." *Asiatische Studien – Études Asiatiques* 73 (3): 521–544.
Rajadhyaksha, Ashish. 1987. "The Phalke Era: Conflict of Traditional Form and Modern Technology." *Journal of Arts & Ideas* 14 (15): 47–78.
Sawhney, Rashmi Devi. 2023. "'Uncontained' Archives of Cinema." In *Exploring Past Images in a Digital Age: Reinventing the Archive*, edited by Nezih Erdoğan and Ebru Kayaalp, 215–234. Amsterdam: Amsterdam University Press.
Spivak, Gayatri Chakravorty. 1994. "Can the Subaltern Speak?" In *Colonial Discourse and Post-Colonial Theory: A Reader*, edited by Patrick Williams and Laura Chrisman, 66–111, New York: Harvester.
Thomas, Rosie. 2014. *Bombay Before Bollywood: Film City Fantasies*. New Delhi: Orient Blackswan.
Thompson, Kristin. 1986. *Exporting Entertainment: America in the World Film Market, 1907–1934*. London: British Film Institute.
Vitali, Valentina. 2008. *Hindi Action Cinema: Industries, Narratives, Bodies*. Bloomington: Indiana University Press.

3 Willful Aesthetics

Pedagogies of Exposure in Animated Short Film

Libe García Zarranz

Introduction: Responding to Film Exposure and Habituation

Let me open this chapter by gesturing toward some of the compelling questions that have guided our collective discussion around cinema *in* and *beyond* the Anthropocene, as proposed by the editors of this volume. How can cinematic forms help us re-situate ourselves in the so-called Anthropocene? And how do these respond to our current climatic, social, or cultural reality? The two central words for me here are the verbs *re-situate* and *respond*, as they both signal an attentiveness to action, change, and ethics. Reflecting on these two questions takes me back in time and space to Zaragoza, Spain, in the 1980s and 1990s. My childhood memories, especially the winters, are often blurry, but there is a consistent sequence of images that remains. Every night, my family and I chose a film to watch. I would go to my bedroom, where over 2,000 VHS films were kept, and collect the selected pick. This was an act of repetition and habituation; an everyday routine that made me feel good. Affect theorist Carolyn Pedwell reminds us that "when affective responses become routine, they can lose their force and may actually prevent meaningful action and change" (2021, 12). Instead of centering major revolutionary moments, Pedwell insists on the power of 'minor' affective responses to generate sociopolitical transformation. It is the "everyday gestures, affects, and habits" (167) that reverberate in Pedwell's work and that resonate for me, as these invite a reconsideration of how the ordinary and the quotidian are deeply enmeshed with social and political potential. As a child and then a teenager, being overexposed to film animated me in paradoxical ways: it comforted me while blinding me to a family dynamic that was saturated by a sea of complicated affects. In Spinozean fashion, the habit of film paralyzed me instead of prompting me to act and seek change. My experience further resonates with Monica Michlin's insights when she contends how overexposure in film "simultaneously reveals illusion and blinds us to it" (2014). While reflecting on these blinding effects, I want to consider instead how overexposure *to* and as illustrated *in* different cinematic forms could plant a seed for action. What if being habituated into certain types of aesthetics fostered a sense of willfulness and agency with repercussions for

DOI: 10.4324/9781003468028-4

individual and social transformation? What if exposure to genres such as animation could help promote ethico-affective pedagogies that are relational, engaged, and collective? This space of the 'what if' is a site of speculation and dreaming, because as film director Djibril Diop Mambéty (Ukadike and Mambéty 1998, 151) reminds us, what is cinema, after all, but magic in the service of dreams.

Willful Aesthetics

Writing from a US educational context, critical literacies scholar Bessie J. Dernikos (2018) conceptualizes classroom environments and texts as charged with affective forces that circulate, often unevenly. These forces, Dernikos claims, position students "in ways that inflict trauma" (2018, 2), which is understood as habitual everyday exposure to restrictive norms around race and gender. In turn, and as a reparative act, Dernikos urges a pedagogy of exposure that "seeks to not only *expose* but also *recover* traumatic wounds by reimagining an affective, albeit risky, relationship to past and present histories of violence" (3). This chapter seeks to translate Dernikos's formulation into the world of animated short film and ask how this subgenre may enact a pedagogy of exposure through what I call 'willful aesthetics' (Herhuth, 2019; Ahmed, 2014). I here put Jonathan Wright's *The Bear Facts* (Nunavut Animation Lab, 2010) in conversation with Aphton Corbin's *Twenty Something* (Pixar, 2021), two animated shorts that unravel ongoing processes of colonialism, while centering Indigenous knowledges and Black kinship respectively. In my view, *The Bear Facts* and *Twenty Something* expose viewers to a 'willful aesthetics' conveyed through the use of various techniques, such as watercolor landscaping, 2D animation, lighting, sound effects, and character animation. In doing so, these cinematic forms become willful objects that respond affectively, ethically, and aesthetically to historical and current climatic, social, and cultural realities. At the same time, and thinking with transdisciplinary geographer Kathryn Yusoff, this re-centering of Indigenous epistemologies and Black collectives counters the exclusionary and extractive nature of anthropocentric discourses and practices. As Yusoff claims, "the Anthropocene is configured in a future tense rather than in recognition of the extinctions already undergone by black and indigenous peoples" (2019, 51). In turn, the animated shorts discussed in this chapter sustain a willful aesthetics where Indigeneity and Blackness persevere.

Formally, if we consider animation as moving images, animated film is then intrinsically a willful genre (Lamarre 2009, ix). However, my formulation of 'willful aesthetics' attends to certain specificities. At the level of filmmaking, I examine animated short films that employ techniques that mobilize sensory and affective realms with the intention of activating ways of re-seeing or recovering erased histories. Following film studies scholar Aylish Wood, I am particularly interested in those animated shorts that "re-animate space" (2006, 135), that is films where space exists as a willful entity that contains experiential meaning and that visibly re-forms. In terms of characterization, the animated short films I have examined also portray insurgent human and more-than-human subjects and environments that respond to their reality in willful ways and are not punished in the process. I here draw from feminist anti-racist scholar Sara Ahmed's conceptualization of willfulness as "a diagnosis of the failure to comply with those whose authority is given" (2014, 1). At the level of reception, I am further intrigued at how animation's willful aesthetics may mobilize a pedagogy of exposure that not only seeks to recover but also respond to ongoing colonial and anti-Black necropolitical forces.

Anishinaabe (Temagami) film and media scholar Tyson Stewart refers to Zacharias Kunuk's *Atanarjuat: The Fast Runner* (Isuma, 2001), the first feature film written, directed, and acted entirely in the Inuktitut language, together with the Truth and Reconciliation Commission (TRC) in 2008, as two of the cultural and political phenomena that have contributed to "a major swell in Indigenous dramatic filmmaking in Canada" (2021, 165). Feature films such as Marie-Hélène Cousineau and Madeline Ivalu's *Before Tomorrow* (Isuma, 2008), Zacharias Kunuk and Natar Ungalaaq's *Searchers* (Kinguliit, 2016), Elle-Máijá Tailfeathers and Kathleen Hepburn's *The Body Remembers When the World Broke Open* (Experimental Forest Films and Violator Films, 2019), and Jeff Barnaby's *Blood Quantum* (Prospector Fims, 2019) constitute what Stewart identifies as a "New Wave of dramatic filmmaking [that] dedicates its focus specifically to Indigenous cultural resurgence" (2021, 165). Following Stewart, I would argue that Indigenous animation in settler Canada, one of the artistic sites I engage with in this chapter, has further revolutionized the genre's aesthetics and purpose, particularly in the last fifteen years, through its innovative techniques and its resistance to reproducing damage-centered storylines (Tuck, 2009). The National Film Board of Canada, for example, has collaborated with the Aboriginal Peoples Television Network (APTN) in several initiatives. These include the multi-episode stop-motion animation series *Wapos Bay* (2006–2011), co-produced by Métis/Saulteaux filmmaker Melanie Jackson; *Vistas* (2009), a collection of 13 films created by Indigenous artists including Jackson's stop-motion short *Dancers of the Grass*; or the CBC Arts series *Keep Calm and Decolonize*, curated by Jesse Wente in 2017, which included Amanda Strong's *Flood*, made with shadow puppetry and stop motion animation.

In the next part of this chapter, I analyze and discuss the Nunavut Animation Lab's 2010 animated short film *The Bear Facts*, produced by the National Film Board of Canada and the Inuit Broadcasting Corporation in association with the Government of Nunavut and the APTN.[1] The short was written, directed, and animated by Jonathan Wright, who grew up in Toronto and moved to Iqaluit in 2005, following his partner, the award-winning Inuk writer-director Alethea Arnaquq-Baril. Wright and several other artists who participated in the National Film Board of Canada's Nunavut Animation Lab workshop in 2008 were chosen to create their own short films. *The Bear Facts*, which is available for free streaming, was one of the animations that resulted from this program ("Artic Edge" 2014). At the level of plot, this four-minute animated short film[2] depicts a somehow humorous encounter between a European colonial explorer and an Inuk hunter which results in a failed attempt at colonization.

In a recent personal communication, Wright (2022) explained how animating the short film was a slow process, as he did frame-by-frame drawings, much like traditional hand-drawn animation. He used Adobe After Effects (AE) for all the animation and drew everything in Photoshop. All the background textures were done traditionally on paper with watercolor. As with other Indigenous animated short films, textures become a narrative tool in *The Bear Facts* that contributes to its willful aesthetics. The textures of the landscape, including the ocean, the sky, and the Artic ice, prioritize the sensory realm and emphasize distinct aspects of Inuit land. As a result, space in this animated short does not only serve to support the actions of the two human characters, but it is drawn as an animated entity in itself. As Wood contends, it is useful "to think about space when it is *not* used-up, when it emerges in an abundance that escapes the setting of place, or the control of character" (2006, 137). In Wright's film, the land acts, changes, and has a will. In this sense, the land's willfulness is embodied and situated; it comes from the territory. As seen in Figures 3.1 and 3.2, the colonizer attempts to claim ownership of the land, saturating the space with an exaggerated number of flags while naïvely protecting himself under a sentry post-shaped armor that gives the illusion of protection and territorial authority. However, the wind begins to blow, turning into a snowstorm, which forcefully removes his shield, leaving him bent and exposed. These natural elements thus willfully align to refuse being conquered. In doing so, the short film resists anthropocentric impulses, such as depicting the land as a site for human extraction, to highlight instead that Inuit territories and ecosystems are agentive. *The Bear Facts* makes the land knowledgeable, capable of making meaning that emerges "through a compassionate web of interdependent relationships that are different and valuable because of that difference" (Simpson 2014, 11). In contrast, the figure of the colonizer is presented as an uncivilized individual lacking local knowledge, which subverts the colonial imperative of making whiteness a marker of civility.

Figures 3.1 and 3.2 Still images from *The Bear Facts*.

The short film contains no dialogue, a strategy that further intensifies the land's sense of aliveness conveyed through the diegetic sounds of the ecosystems drawn through sketching and the use of watercolor landscaping. The insurgent more-than-human soundscapes that reverberate from the animated short thus constitute an integral part of the film's willful aesthetics. The ocean's waves, the clicking of the reindeer, and the wind's force counter the strident metallic noises of the colonizer. The subtle sound of the Inuk hunter's steps, together with his slow-paced movements, further conveys a calm mood, which challenges stereotyped representations of Indigenous peoples in film history as criminal and emotionally excessive.[3] The non-diegetic music of the soundtrack in the film intensifies this soothing feeling, acting like a lullaby that conveys an anticolonial story without violence.

In terms of lighting, the pervasive use of white colors in the animations has a triple function. It exposes viewers to climatic forces such as snow, ice, and daylight, which are marked in the film as willful entities. At the same time, whiteness connotes the overwhelming presence of the colonizer, embodied in this short in the form of a colonial explorer who fails in his attempt to conquer Indigenous land. Light, often associated with knowledge, is further connected to the Inuk hunter, which reverses preconceived colonialist ideas of Indigenous peoples being the object and not the subject of knowledge. This hunter knows all the bare facts and tricks to survive and resist, whereas the colonizer lacks knowledge of the territory which prompts him to fail in his quest. As Ahmed claims, willfulness "involves persistence in the face of having been brought down … Mere persistence can be an act of disobedience" (2014, 2). The Inuk hunter in the film persists by going over his routines, his habits, disregarding the authority of the external force threatening his survival. Habit, according to Pedwell, "is a more-than-human technology which generates both continuity and change; both being and becoming" (2021, 24). Pedwell refers to how digital platforms and other infrastructures also produce habits "in ways that far exceed human intentionality or control" (2021, 170). This extends to external forces and environmental conditions. Circling back to Wright's animated short, the processes of habituation displayed by the hunter in *The Bear Facts* are land-based. He is and becomes *with* the land, forging alliances with her against the colonial threat. Running away in fear, the thin figure of the colonizer fuses with the background textures and vanishes. A few seconds later, the light dims as Indigenous order is restored.

The vitalism of the land here enacts a form of pedagogy that resonates with the work of Michi Saagiig Nishnaabeg scholar and artist Leanne Betasamosake Simpson in that the Inuk hunter learns "*from* the land and *with* the land" (2014, 7). In contrast to the "assimilative colonial agenda" she received when she was a child, Simpson proposes a different approach to education grounded on observation, learning with more-than-human worlds, creativity, and new technologies. This mode of processual learning, which foregrounds Nishnaabeg intelligence, is mirrored in *The Bear Facts* in that the hunter practices Inuit intelligence as insurgent anti-colonial pedagogy. The film's willful aesthetics thus recover and re-situate knowledges that have historically been rendered unworthy while rewriting colonial history from an Indigenous point of view. In this sense, the film further enacts a pedagogy of exposure, in Dernikos's sense, while aligning with other Indigenous animated films that offer a critique of anthropocentric impulses, such as a human-centered focus on the individual and the assumption that land exists as a source for human consumption, in commercial animation.

Walt Disney Animated Studios, in contrast, has been long criticized for its normative lens when it comes to portraying historical processes and present social realities. Disney's extractive tendencies, including the appropriation of Indigenous cultures, the privileging of Western ethnocentrism, and the domestication of nature on screen, have been widely theorized (Mitman, 2009; Giroux and Pollock, 2010). Turning to Pixar Animation Studios, which was acquired by Disney in 2006, we can observe similar patterns in terms of the limits of representation of neurodivergence, together with gender and racial differences, particularly until 2011. Moving into the last decade, some of Pixar's feature films and shorts have started to destabilize these extractive practices by using point of view and other visual techniques in intersectional and non-ableist ways. In *Pixar and the Aesthetic Imagination*, film scholar Eric Herhuth claims that "Pixar's animated worlds ... depict changes to modes of knowing the world, and they depict the reconfiguration of communities" (2017, 10). Herhuth's remarks resonate with Pixar's experimental series SparkShorts (2019-present), particularly in the ways in which these shorts portray a variety of willful subjects whose actions activate ethical ways of collective being and feeling. This initiative, which platforms new storytellers and storytelling techniques, allows invited members of the Pixar team to make a short film in six months with a limited budget.

In the second part of this chapter, I will turn to *Twenty Something*, an eight-minute SparkShort directed by Aphton Corbin, who is a first-time director with ample experience as a story artist in other Pixar titles such as *Toy Story 4* (2019) and *Soul* (2020). In a recent interview, Corbin explains her passion for animation as follows: "I got into animation because I loved drawings, and seeing those drawings come to life [is a] super special moment" (quoted in Potter 2021, para. 3). Her words here already point to the idea of animation's intrinsic willfulness. In his analysis of different drawing movements in animation, media studies scholar Thomas Lamarre proposes to re-see animation through the decoding of movement, contending how "to compose a movement, animation must first decompose or decode it" (2002, 333). The act of recoding and decoding, in his view, exceeds reproducing or imitating live-action cinema. Instead, decoding "reaches into so-called live action and unravels it. It thus goes to the heart of what is 'live'" (Lamarre 2002, 333). Animation's aliveness through movement, at the level of genre, then provides ample room for a discussion of willfulness. As Herhuth argues, animation stands as "a privileged site for analyzing the human experience of personal freedom or sovereign will" (2019, para. 1). The history of character animation, Herhuth further explains, has insisted on the portrayal of agency as artifice through the use of deliberate techniques of manipulation and fabrication. I agree with Herhuth that this is a problematic genealogy, since it presents human experience as disembodied, apolitical, and separated from the world (2019, para. 3). In contrast, shorts like *Twenty Something* animate the characters' willfulness through an emphasis on embodiment, relationality, and the material spacescapes they occupy in ways that lead to action.

Conceived and released in the middle of the COVID-19 global pandemic, *Twenty Something* is set in a nightclub in California some time in a present without COVID-related markers. The film introduces Gia, a young Black woman turning 21 who is about to celebrate entering adulthood with her sister, Nicole. Stylistically, the short is done in 2D, a decision that Corbin made after feeling inspired by other SparkShorts such as *Burrow* (2020), together with her own passion for drawing and for vintage cartoons from the 1970s. Even though Pixar has been celebrated for its pioneer work with 3D animation, the resources and time needed for this technique can be a barrier for those artists with limited budgets, such as those leading the SparkShorts series. It is thus not surprising to note the renaissance of 2D animation in the last few years, linked to how it allows for experimentation and creativity, while foregrounding "compelling stories" (Stefyn 2019), which are some of Pixar's trademarks.

In contrast to 3D, the 2D animations in *Twenty Something* underscore the dimensionality of space. In doing so, they evoke what Wood refers to as "another kind of intensive experience of space, where we are confronted with a perspective that makes a familiar space more uncertain" (2006, 144). The space of the club is thus re-animated, becoming a willful location; in other words, the location does not just serve the purposes of narrative or character development but carries its own history. A discussion of the racial politics of clubbing in the US exceeds the boundaries of this chapter, but it is important to mention that 'nightclub racism,' with its discriminatory door policies, continues to be a prevalent issue. At the same time, clubs have historically been places of counterculture and subversion for Black and other minoritized peoples.[4] This ambivalence is tangible in *Twenty Something*. Significantly, Corbin discusses her choice of location as follows: "Given *my* more introverted personality, the awkwardness of going out to the club felt so relatable and a better location to tell the story" (quoted in Potter, para. 2). Notice how the noun "awkwardness" signals the discomfort of being and feeling exposed. Instead of retracing, however, Corbin stays with the discomfort and exposes both Gia and viewers simultaneously. In Ahmed's (2014) sense, Corbin's persistence could be read as an act of willfulness at the direction level.

Turning to character animation, Gia is drawn wearing a trench coat that hides a baby, a kid, and a teenager, honoring a common trope from slapstick comedy. Corbin explains how the 16-year-old girl symbolizes the superego, the 10-year-old child represents the ego, and the one-year-old baby stands for the id, in psychoanalytic terms. What interests me for the purposes of this chapter, nonetheless, is rather the centrality of the characters' willfulness in this SparkShort. Gia feels very insecure about turning 21 and what that means, hence the idea for Corbin to split her main character into three distinct subjects. And yet, it is the collective effort and sense of kinship between these three girls that ultimately help Gia conquer her anxiety and embrace the mess that is being twenty something. As seen in Figures 3.3 and 3.4, the abundance of medium and long shots of the three characters conveys how willfulness is drawn relationally. This technique challenges the cinema of isolation by presenting a group of Black girls and women together, across ages, literally arm in arm.

Ahmed reminds us that arms "are shaped by history; arms make history flesh. … Intersectionality is arm. Intersectionality is army" (2017, 234). As Figure 3.5 shows, there are various close ups of the girls' arms in the short, showing how they are literally holding each other up, distributing will across age boundaries. In doing so, the three succeed in disturbing some of the expectations of the place, including the possibility of heterosexual romance. At one point in the short, the baby girl throws up on a young man who is approaching them. It could be argued that the one-year-old is already a willful feminist killjoy. Following Ahmed (2010), being a feminist killjoy entails becoming an obstacle or an interruption to episodes of sexist exchange, racial discrimination, or compulsory heteronormativity. These instances of disobedience disrupt the system momentarily. Gia, drawn as this assemblage of Black girls and young women, is in fact forging her own trajectory. In this sense, Gia's initial disorientation becomes a necessary rite of passage into willfulness.

At an aesthetic level, I also consider how lighting and sound contribute to the short film's willful aesthetics. Michlin contends how "different forms of light come imbued with a cultural history" (2014, para. 2), and this certainly applies to the complex historical and affective significance of darkness in relation to Black and other racially marginalized bodies, as previously discussed in relation to *The Bear Facts*. In *Twenty Something*, viewers follow Gia into a nightclub, a dark enclosed space populated by a wide variety of young people with different body types, clothing, gender expressions, and racial backgrounds. The diegetic sound of disco music further contributes to

Figures 3.3, 3.4, and 3.5 Still images from *Twenty Something*.

the ambivalent tone of the film, since it accompanies Gia's feelings of discomfort and being overwhelmed in this space, while reproducing a certain pattern and rhythm that attracts her curiosity and activates her sense of wonder. In other words, Gia's exposure to this ecosystem helps her become a willful subject. The nightclub and its sounds ultimately become a location of possibility for Black girlhood and womanhood to exist and thrive in community. It is also through the mentoring voice of the older sister, Nicole, that Gia gains the will to act. This is another example of how the animated short decenters the imperative of whiteness while inviting audiences to consider how will is distributed in collective ways, shaping bodies and spaces in the process. This will, to borrow Ahmed's words, is Black feminist worlding (2014, 146).

The film closes with the non-diegetic sound of a love song by the retro-soul artist ASTU, which creates a feeling of collective attunement. At the same time, integrating ASTU's voice into the narrative consolidates the film's willful aesthetics by insisting on centering Black subjects in the narrative. Blackness perseveres, "keeps coming up" (Ahmed 2014, 2), thus challenging the paradox of anthropocentric structures which are built through the simultaneous erasure and extraction of minoritized populations and knowledges. Animators, Lamarre explains (2002, 334), often deliberately draw inspiration from or cite traditional art styles. The director of *Twenty Something* explains how she was inspired by Black girls' comics and the desire for more representation of Blackness in animation. In related ways, Jonathan Wright has also continued to animate and illustrate using sketching, collage, and other art styles, in collaboration with Unikkaat Studios, a production company in Iqaluit, which produces Inuktitut-language films. This studio is run by Arnaquq-Barilan, known for animated shorts such as *Lumaajuuq* (2010), which could be included into this ongoing archive of willful animations. Both Corbin and Wright are therefore contributing to the reanimation of Black and Indigenous cultural production by insisting on the potential of animated short film to refuse anthropocentric imperatives while re-fusing this genre with a willful aesthetics that is collective, relational, and firmly situated.

Conclusion

Pedwell's central methodological questions urge a reconsideration of the complex entanglements of affect, habit, and change: "How do images move us (or not), and how does this resultant sensation drive forward wider personal or social transformation?" (2021, 120). In these concluding remarks, I propose to continue thinking with Pedwell's suggestive question, while circling back to my opening remarks on exposure and change through animated film. In my current position as a tenured faculty member in a department of teacher education, I constantly question the habits that are perpetuated in the classrooms, looking for alternative resources to expose myself and the students to other types of visual worlds that have clear ethical and social justice impulses. Animation that centers Indigenous knowledges and Black kinship across human and more-than-human worlds can resist anthropocentric imperatives that insist on centering whiteness as the norm at the expense of minoritized epistemologies. In contrast, animated shorts such as the ones discussed in this chapter can revitalize cinematic space (Wood 2006, 135), opening up for the emergence of a willful aesthetics worth unraveling. As Yusoff contends, if

> the Anthropocene is viewed as a resurrection of the impulse to re-establish humanism in all its exclusionary terms of universality, then any critical theory that does not work with and alongside black and indigenous studies (rather than in an extractive or supplementary mode) will fail to deliver an epochal shift at all.
>
> (2019, 18)

I believe this argument extends to the fields of film studies and pedagogy. Hence, I am confident that learning and unlearning with Black and Indigenous-made media art, including film, video, audio, and digital media, can further help educators and students in Norway and abroad forge new willful habits for social, ecological, and ethical change.

Notes

1 Other companies in Nunavut, in addition to the Nunavut Animation Lab, include Taqqut Productions Inc., an Inuit-owned film production company located in Iqaluit.
2 *The Bear Facts* can be accessed online here: https://www.nfb.ca/film/nunavut_animation_lab_the_bear_facts/.
3 Film studies scholar Martin Berny (2020) discusses what he calls the double stereotype that characterizes literary and film representations of Indigenous Peoples. As Berny (2020, para. 1, 44) contends, Indigenous figures are often depicted as either noble and romanticized or as bloodthirsty and savage. *The Bear Facts* breaks away from this pernicious binary by portraying the Inuk hunter as patient, resourceful, knowledgeable, and capable of being humorous in the face of crisis.
4 For a discussion of house music culture and the tradition of queer of color critique, see Micah Salkind's *Do You Remember House? Queer of Color Undergrounds in Post-Soul Chicago* (2019). See also the touring exhibition *Night Fever: Designing Club Culture 1960 to Today* (2018–2022), co-curated by Cat Rossi, Jochen Eisenbrand, and Nina Serulus, for a broad examination of the design history of the nightclub.

Works Cited

"Artic Edge." 2014. *Applied Arts*. September 2, 2014. https://www.appliedartsmag.com/blog/arctic-edge-a79/.
Ahmed, Sara. 2010. "Killing Joy: Feminism and the History of Happiness." *Signs* 35 (3): 571–592. https://doi.org/10.1086/648513.

Ahmed, Sara. 2014. *Willful Subjects*. Durham and London: Duke University Press.
Ahmed, Sara. 2017. *Living a Feminist Life*. Durham and London: Duke University Press.
Berny, Martin. 2020. "The Hollywood Indian Stereotype: The Cinematic Othering and Assimilation of Native Americans at the Turn of the 20th Century." *Angles: New Perspectives on the Anglophone World* 10. https://doi.org/10.4000/angles.331.
Dernikos, Bessie P. 2018. "'It's like You Don't Want to Read It Again': Exploring Affects, Trauma and 'Willful' Literacies." *Journal of Early Childhood Literacy*, 1–32. https://doi.org/10.1177/1468798418756187.
Giroux, Henry A., and Grace Pollock. 2010. *The Mouse That Roared: Disney and the End of Innocence*. Lanham, Maryland: Rowman & Littlefield Publishers.
Herhuth, Eric. 2019. "The Animated Will: Intelligent Movement and Embodied Intentionality in Character Animation." *Animation Studies* 14. https://journal.animationstudies.org/eric-herhuth-the-animated-will-intelligent-movement-and-embodied-intentionality-in-character-animation/.
Herhuth, Eric. 2017. *Pixar and the Aesthetic Imagination: Animation, Storytelling, and Digital Culture*. Oakland, CA: University of California Press.
Lamarre, Thomas. 2009. *The Anime Machine: A Media Theory of Animation*. Minneapolis: University of Minnesota Press.
Lamarre, Thomas. 2002. "From Animation to Anime: Drawing Movements and Moving Drawings." *Japan Forum* 14 (2): 329–367. https://doi.org/10.1080/09555800220136400.
Michlin, Monica. 2014. "Open Your Eyes Wider: Overexposure in Contemporary American Film and TV Series." *Sillages Critiques* 17. https://doi.org/10.4000/sillagescritiques.3718.
Mitman, Gregg. 2009. *Reel Nature: America's Romance with Wildlife on Film*. Seattle: University of Washington Press.
Pedwell, Carolyn. 2021. *Revolutionary Routines: The Habits of Social Transformation*. Montreal and Kingston: McGill-Queen's University Press.
Potter, Courtney. 2021. "Director Aphton Corbin Talks Pixar SparkShorts' *Twenty Something*." *D23: The Official Disney Fan Club*. September 8, 2021. https://d23.com/director-aphton-corbin-talks-pixar-sparkshorts-twenty-something/.
Simpson, Leanne Betasamosake. 2014. "Land as Pedagogy: Nishnaabeg Intelligence and Rebellious Transformation." *Decolonization: Indigeneity, Education & Society* 3 (3): 1–25.
Stefyn, Nadia. 2019. *CG Spectrum*. March 9, 2019. https://www.cgspectrum.com/blog/3-reasons-to-get-into-2d-animation.
Stewart, Tyson. 2021. "Truth and Reconciliation Cinema: An Ethico-Political Study of Residential School Imagery in Contemporary Indigenous Film." *AlterNative: An International Journal of Indigenous Peoples* 17 (2): 165–174.
Tuck, Eve. 2009. "Suspending Damage: A Letter to Communities." *Harvard Educational Review* 79 (3): 409–427.
Ukadike, N. Frank, and Djibril Diop Mambéty. 1998. "The Hyena's Last Laugh." *Transition* 78: 136–53. https://doi.org/10.2307/2903181.
Wood, Aylish. 2006. "Re-Animating Space." *Animation* 1 (2): 133–152. https://doi.org/10.1177/1746847706068896.
Wright, Jonathan. 2022. Personal communication. Email correspondence with the author. June 15, 2022.
Yusoff, Kathryn. 2019. *A Billion Black Anthropocenes or None*. Minneapolis: University of Minnesota Press.

4 Envisioning Intergenerational Justice

Hope, Despair, and Transformative Action in Climate Change Films

Alexa Weik von Mossner

"People always ask me if I really have hope," states primatologist Jane Goodall in Elizabeth Leiter and Kim Woodard's documentary, *Jane Goodall – The Hope* (United States, 2020). It is not an unreasonable question to ask, given that hope is mentioned in the titles of no less than four of Goodall's books.[1] However, it is worth paying attention to her answer: "My reasons for hope," she explains, "are the young people." This makes sense within the narrative framework of Leiter and Woodard's film, which showcases Goodall's scientific and activist work as an inspiration for younger generations who look at her "like a deity." And indeed, her "relentless commitment and determination to spread the message of hope" speaks to a dire need for visions of the future that are not as bleak as the ones we regularly find in the reports of the Intergovernmental Panel on Climate Change and the Intergovernmental Science-Policy Platform on Biodiversity and Ecosystem Services.[2] In recent years, there has been a marked trend away from the "doom and gloom" of environmental decline and disaster as we see a "sea change" toward more hopeful storytelling (Paoletta 2023). The idea behind this shift is that negative emotions are paralyzing whereas positive ones can propel us forward (Engelhaupt 2017). "My job is to go around and get people to take action," Goodall explains in the film. "If you want someone to change their mind, it's no good arguing. You have to reach the heart." This is what she attempts to achieve, driven by her own hope that when she is "gone, there will be hundreds and hundreds of young people to take over."

As consoling and inspiring as that thought might be, some prominent younger activists have rejected the idea that older generations put their hopes in them. "You all come to us young people for hope. How dare you?" Greta Thunberg famously asked at the 2019 UN Climate Action Summit in New York, the implication being that doing so is a cop-out, a way of deferring responsibility onto those who have not caused the crisis but are supposed to solve it somehow. While it is unlikely that Thunberg was responding to Goodall directly, it is elucidating to consider climate change documentaries in which her UN speech figures prominently. Films such as Nathan Grossman's *I Am Greta* (Sweden/United Kingdom/United States/Germany, 2020) and

DOI: 10.4324/9781003468028-5

Jim Rakete's *Now* (Germany, 2021) aim to give insight into what energizes young activists who fight for a livable future on planet Earth.

In this chapter, I will focus on the narrative strategies of Rakete's film and its investment in critical or "slow hope" and what I call the "courage of despair" as drivers of transformative action. Transformative action has been defined as "levels of action taken to address the causative, inequitable elements and factors perpetuating an identified problem in order to develop and implement solutions" (Jemal and Bussey 2018) or, more succinctly, as "change that becomes sweeping" (Chan 2019). It involves actors who are both subjects of change and agents of change, as is showcased in *Now*. The same is true for the second film I will consider, Dave Davis and Mary Grandelis's *Being the Change* (United States, 2018), which reminds us why it is wrong for older generations to offload the weight of transformative change onto the shoulders of their children and grandchildren. Based on Peter Kalmus's nonfiction book of the same name, *Being the Change* chronicles Kalmus's personal transition to a low-emission lifestyle that involved reducing his carbon footprint by ninety percent. Highlighting that he did this as a prominent, middle-aged NASA climatologist living a "fairly average suburban life," the film argues that it is everyone's responsibility to change their lifestyles, especially if they live in the affluent global North. After briefly introducing my theoretical concepts, I will analyze the films' perspectives on hope, despair, and transformative action. I will argue that they both aim to engage viewers in ways that leave them inspired and hopeful rather than angry or overwhelmed, even though they envision the path to intergenerational justice in somewhat different ways.

Driving Change: Critical Hope, Slow Hope, and Transformative Action

In recent years, hope has been the center of debate in fields as diverse as philosophy (Bloch 1986), psychology (Gallagher and Lopez 2018), sociology (Scribano 2024), history (Mauch 2019), literary studies (Potkay 2022), and pedagogy (Freire 2021). As one might expect, varied definitions of hope exist across these diverse fields of study. According to psychologists Schmid Callina et al., "the bare-bones conception is that hope is a desire for an end perceived to be good and the belief that the end is possible. In this sense, hope is a complex mental state that has an intentional object" (2018, 12). It is worth noting that, in this conception, both hope and desire are not primarily understood as feelings. This focus on cognition and agency echoes philosopher Ernst Bloch's influential definition of hope "not … only as emotion … but more essentially as a directing act of a cognitive kind" (1986, 12).[3] Schmid Callina et al. also emphasize the importance of agency, which they argue separates hope from optimism: "Optimism refers to confidence that a positive outcome will occur but not necessarily to the individual's ability (or perceived ability) to realize a goal" (2018, 15).

Optimism is thus an ingredient of hope, but not the same as it. According to the authors, the (perceived) ability to realize a goal is a defining element of hope.

Agency is also central to a related distinction that has been drawn across many fields between more passive forms of wishful thinking, which lack (perceived) agency and are based on a blind trust in the future on the one hand (Ojala 2017, 78), and "critical" forms of hope on the other hand. "Everybody's life is pervaded by daydreams," writes Bloch in *The Principle of Hope* (1986, 3). "One part of this is just stale, even enervating escapism ... but another part is provocative, is not content just to accept the bad which exists, does not accept renunciation. This other part has hoping at its core and is teachable" (3). In Bloch's understanding, this provocative part of daydreaming is a form of *social dreaming* that has "the power of anticipation" and functions as a potentially transformative "methodical organ for the New, an objective aggregate state for what is coming up" (157). It is fueled by a critical form of hope that does not turn a blind eye to current injustices, current despair, or current crises and catastrophes, rather, it means to recognize these and move toward a better, more just, and sustainable future. According to Tarozzi (2023b, 3–4), critical hope "is characterised by its great transformative power" and "a cognitive and political act of anticipation of what is not yet given" because it acknowledges the status quo, critiques it, and, ultimately, aims at transforming it for the better. With an eye to pedagogy, Tarozzi argues that such critical hope is "what is currently needed, not only to encourage younger generations to overcome the effects of health, economic and environmental crises, but also to avoid succumbing to fatalism in imagining global social justice" (2023a: 52). Perceived agency is again crucial here as a driver of transformative action since without a belief in their ability to affect the situation, the younger generations that Jane Goodall and so many others put their hopes in might not be able to realize their potential because they do not believe that they can.

This is obvious to those of us who are regularly in conversation with these younger generations, be it at home or in classroom settings. Alyssa Hull reported in a 2019 article that many of her "students hear the words we have 18 months to tackle climate change, or it will be too late, and they think that in 18 months they are going to die." Writing only weeks before the outbreak of the COVID-19 pandemic and the millions of lives it claimed, Hull speculated that communicating the urgency of the climate crisis in such alarming terms might inspire her students "to join the youth climate movements" but it might also be "the thought that paralyzes them, anesthetizes them, and keeps them away and keeps them asleep" (2019). Given that she was writing in the same year that also saw Thunberg's speech in front of the UN and the biggest global climate strike in history, this is a remarkable observation that reminds us that not everyone was taking to the streets in September 2019.

Even though empirical evidence suggests that hope appeals are not necessarily more effective than fear appeals when it comes to inspiring action on climate change (Roberts 2017; Ettinger et al. 2021), there seems to be a psychological need and a deep-seated desire for more hopeful visions of the future.

The question is how that need might be addressed. I have written about the fictional envisioning of more sustainable futures elsewhere (Weik von Mossner 2024), but I believe it is crucial to also consider hopeful climate storytelling in the nonfiction realm. Environmental historian Christof Mauch has championed "slow hope" as a way of using storytelling to make it through our global environmental crisis. With reference to Rob Nixon's concept of "slow violence" as "a form of environmental injustice and violence that is dispersed across time and space" (2011, 1), Mauch argues that

> we need stories not only of the "slow violence" of environmental degradation, but also of what I call "slow hope." Rob Nixon has shown us that oil spills, toxic drifts, and climate change have created violence around the globe. He calls it "slow violence" because it is often invisible and develops slowly and gradually. Nixon argues that we should be more attentive to the damaging impact that environmental change has for people who are vulnerable, poor, and disempowered. Nixon's analysis is perceptive and astute. It helps us understand the creeping character of violence. But, I argue, we also need to identify stories, visions, and actions that work quietly towards a more hopeful future.
>
> (2019, 17)

Mauch's "slow hope" is a form of critical hope in that it is conscious of the forms of slow violence that affect communities and their environments around the globe. Among his examples of such stories, visions, and actions that nevertheless work toward a more hopeful future are Tsai Jen-Hui, an architect at National Taipei University of Technology who has "transformed his university into a green oasis" (2019, 21) and a children's initiative, Plant for the Planet, "that since 2007 has advocated the planting of trees around the globe in an effort to reduce the effects of climate change [and] successfully planted more than 15 billion trees by 2018" (26–27). Reminding us of Bloch's insight that "the most tragic form of loss ... is the loss of the capacity to imagine that things could be different" (1986, 71), Mauch calls for identifying more of such "visions and paths that will guide us away from past dependencies and traps and into understandings that imagine a different, a more just and more ecological, world" (2019, 18). He also reminds us that "hope, for Bloch, has its starting point in fear, in uncertainty, and in crisis" (19). It is an important reminder not to be forgotten in a time when any negative emotion in response to the climate crisis is suspected to be counterproductive.

Given the timeliness of the issue and the manifold calls for hopeful storytelling, it is unsurprising that the power and, indeed, *necessity* of reframing environmental narratives from despair to hope has also been raised in environmental impact filmmaking. At a panel discussion entitled "An Antidote to Despair: How Independent Film is Driving Climate Action" that the Redford Center organized at the 2024 Sundance Film Festival, Steven Olikara put the issue as follows, citing empirical research on framing:

> If you frame an issue in terms of how you can take concrete steps to build a more positive future, you are going to see a lot more uptake because people feel there is something that [they] can do, [they] feel that agency. On the flip side, if you frame the issue as just ten-alarm fire it can lead to a lot of paralysis, and in fact we saw some negative uptake.
>
> (Redford Center 2024)

Film financer Brian Newman agreed, stating that what audiences, especially younger audiences, "really are tired of is despair. And they want positive stories ... so they can embrace something and feel positive about the future and change" (Redford Center 2024). The Redford Center has championed many such films over the years and supported their ability to have a positive impact on the issues and communities they portray; Grossman's *I am Greta* and Chris Cooper's *Youth v Gov* (United States, 2020) are just two of many examples. To these we could add solution-oriented films produced elsewhere in the world, such as the French documentary *Demain/Tomorrow* (Cyril Dion and Mélanie Laurant, 2015) and the Australian film *2040* (Damon Gameau, 2019), which introduce viewers to existing projects of "slow hope" in cities and communities around the world.[4] Jim Rakete's *Now*, to which I will now turn, is another film that aims to energize and enable audiences by giving them hope. It does so by portraying the efforts of young activists.

Critical Hope and the Courage of Despair: Youth Climate Activism in Jim Rakete's *Now*

Now was released in 2021, two years after the Global Climate Strike and Greta Thunberg's "How dare you?" speech at the UN Climate Action Summit—and in the middle of the COVID-19 pandemic. Known as a famous celebrity photographer in his native Germany and beyond, Rakete embarked on making his very first film at the age of 68 because—fifty years after starting out as a news photographer who documented the student revolts of 1968—he was impressed by the Fridays for Future activists and wanted to document their work in a way that would engage others and further energize the movement. Compared to Grossman's *I am Greta*, which attempts to give an intimate portrait of Thunberg's life and work, and Cooper's *Youth v Gov*, which focuses on the Juliana v. The United States of America constitutional lawsuit,

Rakete's film offers a more international perspective.[5] The main protagonists are German youth activists such as Luisa Neubauer from Fridays for Future, Nike Mahlhaus from Ende Gelände, a civil disobedience movement occupying coal mines, and Felix Finkbeiner who, at the age of 9, started the Plant for the Planet initiative that Mauch considers an example of "slow hope" and who by now is a seasoned activist. The German protagonists are joined by the British activist Zion Lights from Extinction Rebellion and the American Vic Barrett, who is one of the 21 youth who sued the US government and a prominent character also in Cooper's film. *Now*'s international outlook and appeal are also expressed in its language. Although most of the portrayed activists are German, they all speak English during the interview portions, using the *lingua franca* of global youth movements.

What drives and energizes these movements is one of the central questions that are explored in the film. Rakete has said in an interview that he "felt the urge to make a film about ... hope, about the hope of the protestors, about the hope of scientists" (The Impact Series interview 2020). Yet hope is not even mentioned in the film's opening sequence, in which the protagonists one by one answer the question of when they first realized they were activists. Rather, it is concrete actions that defined that important moment for them, like quitting a job, planting trees, trying to get their parents to recycle, blocking a coal mine, or disrupting a fossil fuel panel in the US Congress. If they bring up emotions at all, it is a sense of excitement and fear or a feeling of empowerment resulting from communal action. Nike Mahlhaus remembers that it "was amazing to feel part of so many people" when she first went into a coal mine together with two-thousand other Ende Gelände activists: "We shared this moment [when] we realized we're gonna go from here and we're gonna block the infrastructure." Vic Barrett is even more explicit when he says that "going into the space [of a US Congress panel on fossil fuel] and disrupting it in that way was one of the first times I really felt like I made a difference in a way that was scary and uncomfortable at first but afterwards felt so good."

He smiles when he says this as the film cuts to footage from the Global Climate Strike on September 20, 2023. Young activists are chanting "What do we want? Climate Justice! When do we want it? Now!" Thunberg on stage in New York tells people that "we are a wave of change. Together, and united, we are unstoppable." This is followed by a short clip from Thunberg's speech at the UN three days later. "How dare you continue to look away?" she asks, and the film shows US President Donald Trump getting up from his seat before cutting to a rally in Rochester where he complains that "We need some global warming. It's freezing." Thunberg appears to comment on this mockery with another moment from her UN speech: "If you choose to fail us," she says emphatically, "we will never forgive you." The last words of the opening sequence, however, belong to Barrett, who addresses the New York climate strikers on a stage in Foley Square and leads them in chanting: "This is our time / This is our future / And I will do / Whatever it takes / To protect our lives / To protect our planet / And to protect our people." Once again, the focus is on community and *transformative action* rather than waiting for change.

That the music score accompanying these scenes is energetic, almost triumphant, is a deliberate choice aimed at provoking a visceral response.[6] Rakete has explained that he wants viewers to be "swept away" by the power of the music and the images filled with thousands of young climate activists demanding a livable future (Haus des Dokumentarfilms interview 2021). He originally wanted to make what he calls a "small film"—produced with very few people in a minimal amount of time that would be "chockfull" of arguments. It was only when he realized that the long string of talking heads was overwhelming viewers rather than convincing them that Rakete decided to add more breathing space to the film with images and music that would communicate visually and audibly the compelling force of the climate youth movement. Throughout the film, several montage sequences show the young strikers marching in cities around the world, giving a sense of scale and the "unstoppability" that Thunberg and others have claimed for the climate movement. To provide some historical perspective, Rakete has included interviews with several witnesses of the protest movements of the 1960s and 1970s, who share their thoughts on why those earlier movements faltered, and why the circumstances make this new moment of protest and activism different and much more promising. "For the first time in my life, I feel that this is not gonna stop," Rakete shares in the Impact Series interview, "This is going to go on." This optimism pervades every minute of the film, the sense that this global movement might be the first one that can actually transform societies.

The current state of research supports this stance. A 2023 study conducted by Dylan Bugden (2020) shows that the peaceful marches of the Fridays for Future movement had a stronger positive effect on shifting public support for climate change action than civil disobedience or other activist tactics.[7] However, it is important to note that positive emotions and thought patterns such as hope and optimism are not the only drivers of transformative action. "I think Extinction Rebellion does a very good job at finding a place for emotions," says Nike Malhaus in Rakete's film and goes on to explain that

> we from Ende Gelände ... have always spread hope and optimism. We have always said "We can still change something if we act now, we can prevent climate crisis from becoming a catastrophe." And in the last year or so, I realized, that this is not how I feel any longer. I actually feel desperate quite a lot. And I think that Extinction Rebellion are better at taking up this emotion ... these negative emotions also. Climate anxiety. Really a deep anxiety of not having a future. I like their slogan, they say "Hope dies. Action begins." And I think we could do better at finding places for these negative emotions. Because they grow. They grow, and we need to turn them into good feelings that enable us to act and don't prevent us from acting.

It is a remarkable and powerful admission in a film that is supposed to spread hope, and yet it is echoed in Thunberg's "How Dare You?" speech before the UN General Assembly, which, together with the climate protest marches in New York and other metropolises around the world form *Now*'s dramatic climax. Thunberg's speech is given almost in its entirety, but Rakete chooses to present most of it as voiceover while viewers once again see the marching climate strikers in the streets of Berlin, London, and New York.

Thunberg's angry rejection of older generations "coming to us young people for hope" is a powerful reminder that outrage can also be an important driver of transformative action. Reporting the results of their study on emotional responses to climate change communication, Teresa Myers and colleagues conclude that

> it may be especially important to communicate these messages in combination, communicating not only the hope we have for moving forward on climate change but also the reasons such actions are imperative, or conversely, which are not only reasons for concern but also paths for action.
> (2023, 9; see also Featherstone 2023)

At times, hope might not be necessary at all to clear those paths forward. In *I Want a Better Catastrophe* (2023), Andrew Boyd recounts his gradual loss of hope over the course of decades. He admits that, now in his fifties, he has lost his faith that humanity will be able to address the global crisis it is facing. "I've done the climate math," he explains, "and I see no way forward. I am filled with despair. And then a strange thing happens: I feel fierce. I feel clear. I feel free. I don't give a fuck anymore. I've got nothing left to lose. I'm willing to take risks that I wasn't before" (2023, 74). There is a term in German for this state of mind: *Mut der Verzweiflung*, the courage of despair. "It turns out despair is its own kind of power," writes Boyd, "its own kind of freedom. And then I think: If enough of us fall into a dark enough despair, who knows what we can do together. This is the only hope I have left" (74). The transformative power of the courage of despair is also expressed in the Extinction Rebellion slogan Mahlhaus cites in the film: "Hope dies. Action begins." And yet, *Now* ends on a hopeful note, and it is Mahlhaus who is given the final word. Rather than ending with images of large groups of young people protesting, Rakete chooses to show her pondering the very personal question of whether she wants to have children. "At the moment I think yes," she says, "because it gives me hope. And I think it means that we don't give up ... I want to create a better world, and I want to have children that I can raise into this better world."

It is this combination of personal hopes, intergenerational thinking, and transformative action that has made the demands of the youth climate movement relatable. A 2022 study by Spaiser et al. observes that

> Friday for Future's normative framework makes the shared, unjust casualty experience of young people ... accessible to the public.

> The victims are now in spatial, temporal, and social proximity, they are our children and grandchildren, and this makes the normative challenge of the status quo (continuation of fossil-fuel based economy) so potent. (2022, 1)

Whether that potency will be able to change the status quo, as not only Rakete and Goodall, but also these researchers seem to hope, is an open question. Recent years have seen a marked decline in youth climate activism (Salter 2022), in part because the COVID-19 pandemic made large-scale protest marches impossible and in part because it is difficult to maintain the energy and optimism of protest movements in the face of overwhelming odds.[8]

From Hopeful Visions to Actual Change: Engaged Filmmaking and Transformative Action

Now was released in 2021, into a world that had experienced a near-standstill due to COVID-19. Many observers were staggered not only by the pandemic itself and the loss of life it claimed, but also by the realization that governments and citizens are in fact able to (re)act quickly and comprehensively when they perceive a clear and present lethal danger. As economic activity was reduced to a trickle and commercial planes were grounded for weeks, many were wondering whether it would be the beginning of a new era, a forced break in habits that might help people in the industrialized Global North to adopt a less carbon-spewing lifestyle that is more equitable and ecologically sustainable. By now, researchers have shown that the pandemic years saw a significant increase in the use of private aviation, leading to "an increase of CO_2-equivalent emissions by 23%" rather than a reduction (Sobieralski and Mumbower 2022). Hopes that at least commercial air travel would not climb back to pre-pandemic levels were also thwarted. July 13, 2023 now holds a sad record as the day with the greatest number of planes in the air in human history, breaking a record set only seven days earlier, on July 6 (FlightRadar24). Clearly, the break in bad habits was only temporary.

However, accepting the climate youth movement's normative challenge to the continuation of a fossil-fuel based economy can nevertheless lead to real change on a more individual level. Asked about his own carbon footprint in a 2020 interview with the German magazine *Cicero*, Rakete said that he rarely drives cars anymore and travels sparingly: "I became this gray figure on the bike, in the bus lane" (Boehle 2020; my translation). Talking to the young climate activists made him realize that he would have to change his personal consumption habits because he belongs to the generation that has caused the problem. That such insights lead to personal lifestyle changes is still rare. It is not unusual to hear people speak about their climate anxiety and, in the same conversation, about the long-distance flight they booked for their next vacation. In her book *Living in Denial* (2011), the Norwegian sociologist Kari Marie Norgaard explains this problematic gap between awareness and action in terms of a socially organized denial, by which information about the climate crisis is known in the abstract, and even causes feelings of distress and anxiety, but entirely disconnected from political, social, and private life.

In a conversation with German Fridays for Future activists, the Canadian filmmaker Slater Jewell-Kemker, who directed another documentary about the youth climate movement, *Youth Unstoppable* (Canada, 2021), reflected on the need to not only influence policymaking and public opinion, but to also harmonize one's values with one's personal behavior. Jewell-Kemker was only fifteen years old when she started chronicling the movement that she herself was part of; at the age of twenty-eight she stresses that she is "trying to move beyond just saying things and having made a film," because she believes that even though systemic societal change is what is ultimately needed, "the way that you change yourself and the way that that impacts the people around you and the community is enormous." Peter Kalmus would certainly agree. In his book *Being the Change: Live Well and Spark a Climate Revolution* (2017), the NASA climatologist comes to the same conclusion and shares how he reduced his own carbon footprint to about one-tenth the personal average in the United States. What led him to change his life was a combination of scientific knowledge and that he became a father. Rather than hoping that his children's generation would somehow fix that situation, he changed his own life.

Slow Hope and Voluntary Sufficiency in *Being the Change*

Dave Davis and Mary Grandelis's 56-minute film, *Being the Change*, aims to capture the essence of Kalmus's book, which, as Erin James has noted, is part popular science and part personal memoir. "The rhetorical purpose of *Being the Change*," writes James, "is to immerse readers in that new story and its better world as a means of both overcoming environmental apathy and dealing with climate grief" (2022, 184). This is true for both the book and the film, which opens with archival footage and Kalmus's voiceover as he remembers the first time he heard about global warming. He was in sixth grade at the time, still a child, and the idea of a changing climate "seemed like science fiction" to him. Twenty years later, when he had his first son and was a graduate student in physics at Columbia University, he went to see Davis Guggenheim's climate documentary *An Inconvenient Truth* (United States, 2006), featuring Al Gore. When he left the movie theater, he was "kind of shell shocked. How can there be this incredibly scary thing on one hand, and then on the other hand, the expert on this is suggesting that we should change our light bulbs?" Realizing that this minimal action would hardly be enough to address the problem set Kalmus on a path of action. In the film, he remembers that

> I was feeling very anxious at the time because I had this knowledge of this stuff happening [and] that society wasn't moving to do anything about it. And that gave me a sense of panic. And then I realized that I myself wasn't doing anything either.

It was this realization of his own inaction and complicity that led Kalmus to not only study climate change but also engage in transformational action that would change his own life and that of his family. "Changing ourselves," he explains in the film, "we send a message with how we live that goes far deeper than what we can say with words." To bring his personal life in harmony with his knowledge and values, and thus to "be the change" rather than just hoping for it, he took a hard look at his own carbon footprint.

Realizing that most of his emissions were related to the flights he took to scientific conferences and other work-related trips, Kalmus made the decision that he would stop flying, regardless of the potential professional repercussions. "The quickest way a normal human can warm the planet is by flying on a commercial plane," he writes in *Being the Change* (2017, 151), so there was no way around quitting that. Since his second-largest climate impact came from food, he transitioned to vegetarianism, started gardening, and composts his family's food waste. He also rides his bike to work and makes longer trips in an old car that is powered by discarded vegetable oil. "Overall, this hasn't been a sacrifice," he writes. "It has made [him] happier" (6) because he now lives a life that is aligned with his values. Central to this personal transformation is an approach that Kalmus calls "the head, the hands, and the heart." As he explains in the film, the head stands for an "intellectual understanding of the problem of climate change in its scientific, ecological, and societal dimensions," the hands for "actually doing stuff instead of just talking about things or thinking about things or reading about things." However, it is the third element, the heart, that "ties everything together." The emotional side of things is where, according to Kalmus, "you find meaning in this. It's where you translate this intellectual understanding and this practical action into actually making your life more satisfying." The film aims to show how this personal satisfaction reveals itself, showing Kalmus with his family, working in his garden, riding his bike, or meeting with like-minded colleagues and community members.

In Rakete's terms, *Being the Change* is a "small film," not only because it lacks a sweeping music score or the visual force of global protest marches, but also because it keeps the focus on how the individual can engender transformative action. Its biographical approach is comparable to Grossman's *I Am Greta* in that it is an intimate portrayal of a climate activist and his community, the main difference being that, now in his fifties, Kalmus belongs to one of the generations that have *caused* current levels of global warming. "I do realize that everything I do is just a tiny drop in the bucket," he says at one point in the film. "I try to have an effect on my local community and the people around me, and I know that if there are people everywhere around the whole world doing similar things in their community, then eventually that's how things are starting to shift."

In Mauch's terms, the film tells one of those "stories, visions, and actions that work quietly towards a more hopeful future" (2019, 20) and is thus an example of slow hope. "I hope you'll join me," Kalmus writes at the end of his book, "we can build this new world together" (2017, 296). The book and the film—which is available for free on YouTube—are meant to make it easier for others to take that step. Kalmus also founded No Fly Climate Sci, a website for "scientists, academics, and members of the public who either don't fly or who fly less" because "flying is the domain of the globally privileged" and "actions speak louder than words." Such actions might seem radical in a society like the United States, where people continue to rely on air travel to cross the vast distances between destinations. And yet, they might benefit not only future generations and the planet, but also the person engaging in the action. "Burnout of the Earth's resources and emotional burnout are related to one another," writes Mauch. "In a world of accelerated consumption, resource exhaustion reveals itself to be not only a physical but also a sociological and psychological phenomenon" (17). *Being the Change* (the book and the film) offers advice on how to slow down consumption and discover a better way to live. Its focus on individual direct action highlights perceived and actual agency and the resulting *slow* hope that, according to Mauch, is a driver of change and an antidote to despair.

Conclusion

Climate change films such as *Now* and *Being the Change* deliberately move beyond the communication of scientific findings and the use of apocalyptic imagery. Instead, they showcase movements and individuals who engage in transformative action to address climate change and work toward ensuring more intergenerational justice. They are joined by a slate of other recent solution-oriented films that aim to inspire viewers of all ages and generations to join the climate movement and/or engage in transformative action in their own lives and communities. As environmental documentaries, these films likely reach only small and self-selected audiences in the industrialized Global North, so their success rates might be modest. However, they can be conversation starters and create the slow ripple effects that Kalmus and other climate activists are hoping for, to build broad alliances for intergenerational climate justice. Hopefully, those alliances will grow to include more prominent scientists and scholars.

Despite her tireless engagement for environmental issues, Jane Goodall is not on the No Fly Climate Sci list. A 2017 article in *Scientific American* reported that she travels as much as 300 days per year and that people regularly approach her "in airports, tears in their eyes, and tell her she's their idol" (Weintraub 2017). Not much has changed since then despite Goodall's advanced age. In January 2024, she traveled to the World Economic Forum in Davos, Switzerland to remind people to vote in a year when more than forty countries elect new leaders because "the health of Earth itself is on the ballot" (Gelles 2024). Like Al Gore and scores of other celebrities and scientists who speak up for environmental issues, Goodall appears to operate under the assumption that the good she does with her political work around the globe justifies her carbon footprint. Perhaps this assumption is correct. And yet, doing so contributes to a lifestyle that destroys the future of the generation she puts her hopes in. As Goodall recognizes on her website, "people who have access to wealth and resources often have a disproportionately large carbon footprint based on their consumption habits .... Current and future generations will bear the greatest burdens of climate change, although previous generations have caused it."[9] Given this recognition, what do older generations owe to those current and future generations? Kalmus's choice to quit air travel has been criticized as too radical and career-ending, but at the time of my writing he is still a climate scientist at NASA's Jet Propulsion Laboratory and, together with his other lifestyle changes, it is a choice that has helped him drastically reduce his carbon footprint regardless of his privileged access to wealth and resources. Moreover, he claims that he is living a more fulfilling life. "Some of the best motivation is this internal motivation," explains psychologist Janet Swim in the film *Being the Change*. "Peter, when looking at his own life, felt he needed more alignment," and so he made changes that allowed him to address some of the contradictions.

This is a need I can relate to. As someone who has worked for many years on cinematic representations of climate change, I find it increasingly difficult to separate my scholarly engagement from my personal life. I am not on Kalmus's No Fly Climate Sci list either, but I finally quit planes. My last flight was on July 15, 2023, from Portland, Oregon to Frankfurt, returning from the biennial conference of the Association for the Study of Literature and Environment. The temperature in Portland had been in the 100s, wildfires were devastating parts of California, and Europe was also badly affected by extreme weather. Everyone I talked to at the conference had expressed their worries about accelerating climate change and their great respect for the youth climate movement and its demands. When, after my return, I learned of the two shocking records in commercial aviation set in July 2023 and realized that I had personally contributed to them, I reacted precisely in the way that Kalmus describes in *Being the Change*: I was feeling anxious because I "had this knowledge of this stuff happening [and] that society wasn't moving to do anything about it .... And then I realized that I myself wasn't doing anything either." I watched the film shortly after that and have not booked another flight since. I have also changed my life in other ways to create more alignment between my actions and my values. Like Kalmus, I have found it liberating.

Notes

1 Goodall's books with hope in their title are *Harvest for Hope* (2005), *Hope for Animals and Their World* (2009), *Seeds of Hope*, with Gail Hudson (2013), and *The Book of Hope*, with Douglas Abrams and Gail Hudson (2021).
2 The phrase has been used in marketing material for the film as well as on the website of the Jane Goodall Institute.
3 Psychologist C.R. Snyder has argued that hope is not an emotion at all, but a positive cognitive-motivational state marked by strong agency and motivational determination (2002). This seems to go a bit far, however, since most people will say that they *feel* hopeful and hope is also frequently understood to be the opposite of despair, a strong negative emotion.
4 For an analysis of the two films' narrative strategies, see Weik von Mossner (2022).

5 Juliana v. The United States of America was a lawsuit filed in 2015 by 21 American youth against the US government for violating their constitutional rights to life, liberty, personal safety, and property through their willful actions in creating the climate crisis (https://www.lwv.org/legal-center/juliana-v-united-states-0).
6 It is a scientifically proven fact that film music has a visceral effect on viewer emotions (Smith 1999).
7 Bugden notes, however, that the greater positive effect of peaceful marches could potentially be balanced out by the greater media attention garnered by civil disobedience.
8 The recent crisis faced by youth activism is captured, for example, by the German documentaries *Aufschrei der Jugend* (Kathrin Pitterling, 2021) and *Bis hierhin und wie weiter?* (Felix Maria Bühler, 2023).
9 See https://janegoodall.ca/the-3-crises/climate-change.

Works Cited

Bloch, Ernst. 1986. *The Principle of Hope*. 3 volumes. Translated by N. Plaice, S. Plaice, and P. Knight. Cambridge, MA: MIT Press.
Boyd, Andrew. 2023. *I Want a Better Catastrophe*. Gabriola Island: New Society Publishers.
Bugden, Dylan. 2020. "Does Climate Protest Work? Partisanship, Protest, and Sentiment Pools." *Socius* 6. https://doi.org/10.1177/2378023120925949.
Chan, Kai. 2019. "What Is Transformative Change, and How Do We Achieve It?" *IPBES* Blog, November 11. https://www.ipbes.net/news/what-transformative-change-how-do-we-achieve-it.
Engelhaupt, Erika. 2017. "Does Doom and Gloom Convince Anyone about Climate Change?" *Science News*, July 28. https://www.sciencenews.org/blog/science-public/new-york-magazine-climate-change.
Ettinger, Joshua, Peter Walton, James Painter, and Thomas DiBlasi. 2021. "Climate of Hope or Doom and Gloom? Testing the Climate Change Hope vs. Fear Communications Debate through Online Videos." *Climatic Change* 164: 19.
Featherstone, Liza. 2023. "The Case Against Both Climate Hope and Climate Despair." *New Republic*, July 31. https://newrepublic.com/article/174719/case-climate-hope-climate-despair.
Freire, Paulo. 2021 [1992]. *Pedagogy of Hope: Reliving Pedagogy of the Oppressed*. London: Bloomsbury.
Gallagher, Matthew W., and Shane Lopez, eds. 2018. *The Oxford Handbook of Hope*. New York: Oxford University Press.
Gelles, David. 2024. "Jane Goodall Says 2024 is the 'Most Consequential Voting Year.'" *New York Times Climate Forward Subscriber Newsletter*, January 23. https://www.nytimes.com/2024/01/23/climate/jane-goodall-says-2024-is-the-most-consequential-voting-year.html.
Hull, Alyssa. 2019. "Hopepunk and Solarpunk: On Climate Narratives That Go Beyond the Apocalypse." *Literary Hub*. https://lithub.com/hopepunk-and-solarpunk-on-climate-narratives-that-go-beyond-the-apocalypse/.
Impact Series. 2020. "Interview with *Now* Director Jim Rakete." *Vimeo*, September 28. https://player.vimeo.com/video/458826327.
James, Erin. 2022. *Narrative in the Anthropocene*. Columbus: Ohio State University Press.
Jemal, Alexis, and Sarah Bussey. 2018. "Transformative Action: A Theoretical Framework for Breaking New Ground." *Ejournal of Public Affairs*, August 23. https://www.ejournalofpublicaffairs.org/transformative-action/.

Kalmus, Peter. 2017. *Being the Change: Live Well and Spark a Climate Revolution*. Gabriola Island: New Society Publishers.

Mauch, Christof. 2019. "Slow Hope. Rethinking Ecologies of Crisis and Fear." *RCC Perspectives: Transformations in Environment and Society*. https://doi.org/10.5282/rcc/8556.

Myers, Teresa A, Connie Roser-Renouf, and Edward Maibach. 2023. "Emotional Responses to Climate Change Information and their Effects on Policy Support." *Frontier Climate* 5. https://doi.org/10.3389/fclim.2023.1135450.

Nixon, Rob. 2011. *Slow Violence and the Environmentalism of the Poor*. Cambridge, MA: Harvard University Press.

Norgaard, Kari Marie. 2011. *Living in Denial: Climate Change, Emotions, and Everyday Life*. Cambridge, MA: MIT Press.

Ojala, Maria. 2017. "Hope and Anticipation in Education for a Sustainable Future." *Futures* 94: 76–84. https://doi.org/10.1016/j.futures.2016.10.004.

Paoletta, Kyle. 2023. "The Incredible Disappearing Doomsday." *Harper's Magazine* (April). https://harpers.org/archive/2023/04/the-incredible-disappearing-doomsday-climate-catastrophists-new-york-times-climate-change-coverage/.

Potkay, Adam. 2022. *Hope: A Literary History*. Cambridge: Cambridge University Press.

Redford Center. 2024. "An Antidote to Despair: How Independent Film is Driving Climate Action." https://www.redfordcenter.org/resources/an-antidote-to-despair-how-independent-film-is-driving-climate-action/.

Roberts, David. 2017. "Does Hope Inspire More Action on Climate Change than Fear? We Don't know." *Vox*, December 5. https://www.vox.com/energy-and-environment/2017/12/5/16732772/emotion-climate-change-communication.

Salter, Eleanor. 2022. "Children Aren't the Future: Where Have All the Young Climate Activists Gone?" *The Guardian*, June 29, 2022. https://www.theguardian.com/commentisfree/2022/jun/29/children-arent-the-future-where-have-all-the-young-climate-activists-gone.

Schmid, Callina, Kristina, Nancy Snow, and Elise D. Murray. 2018. "The History of Philosophical and Psychological Perspectives on Hope: Toward Defining Hope for the Science of Positive Human Development." In *The Oxford Handbook of Hope*, edited by Matthew W. Gallagher and Shane Lopez, 9–25. New York: Oxford University Press.

Scribano, Adrian. 2024. "The Sociology of Hope: Classical Sources, Structural Components, Future Agenda." *Society* 61: 1–8.

Smith, Jeff. 1999. "Movie Music as Moving Music: Emotion, Cognition, and the Film Score." In *Passionate Views: Film, Cognition, and Emotion*, edited by Carl Plantinga and Greg M. Smith, 146–67. Baltimore: Johns Hopkins University Press.

Snyder, C. R. 2002. "Hope Theory: Rainbows in the Mind." *Psychological Inquiry* 13 (4): 275. https://doi.org/10.1207/S15327965PLI1304_01.

Sobieralski, Joseph B., and Stacey Mumbower. 2022. "Jet-setting during COVID-19: Environmental Implications of the Pandemic Induced Private Aviation Boom." *Transportation Research Interdisciplinary Perspectives* 13. https://doi.org/10.1016/j.trip.2022.100575.

Spaiser, Viktoria, Nisbett, Nicole, and Cristina G. Stefan. 2022. "'How dare you?' – The Normative Challenge posed by Fridays for Future." *PLoS Climate* 1 (10). https://doi.org/10.1371/journal.pclm.0000053.

Tarozzi, Massimiliano. 2023a. "Futures and Hope of Global Citizenship Education." *International Journal of Development Education and Global Learning* 15 (1): 44–55. https://doi.org/10.14324/IJDEGL.15.1.05.

Tarozzi, Massimiliano. 2023b. "Introduction. Introducing Pedagogy of Hope for Global Social Justice." In *Pedagogy of Hope for Global Social Justice. Sustainable*

Futures for People and the Planet, edited by Massimiliano Tarozzi and Douglas Bourn, 1–10. London: Bloomsbury.

Weik von Mossner, Alexa. 2022. "Climate on Screen: From Doom and Disaster to Ecotopian Visions." In *The Cambridge Companion to Literature and Climate*, edited by Kelly Sultzbach and Adeline Johns-Putra, 162–176. Cambridge: Cambridge University Press.

Weik von Mossner, Alexa. 2024. "Wish We Were There: Hope, Desire, and Utopian Community in Contemporary Solarpunk." *Utopian Studies* 35 (2).

Weintraub, Karen. 2017. "Jane Goodall, Still Traveling the World and Speaking Up for Animals at 83." *Scientific American*, October 9. https://www.scientificamerican.com/article/jane-goodall-still-traveling-the-world-and-speaking-up-for-animals-at-83/.

5 Take Back the Walk

Trekking and Female Empowerment in *Wild* and *Tracks*

Virginia Luzón-Aguado

Introduction

In her book *Affirmative Aesthetics and Wilful Women*, Maud Ceuterick sets out to explore how the "willful" women that appear in the films she analyzes strive to claim such male-dominated spaces as "the street, the house and the car" as their own (2020, 2). They do so both as an act of rebellion against patriarchal restrictions and as an affirmation of their female selves. Meanwhile, Uteng and Cresswell have argued that "acquiring mobility is often analogous to a struggle for acquiring new subjectivity" (2016, 2). Thus, my aim in this chapter is to expand on Ceuterick's argument by proposing the natural outdoors as a space for female mobility, agency, and rebellious affirmation. In order to do this, I will provide a reading of *Wild* (Jean-Marc Vallée, United States, 2014) and *Tracks* (John Curran, Australia 2013). The films are based on the true stories of Sheryl Strayed and Robyn Davidson, respectively, who undertook major trekking expeditions on their own. By challenging long-established gender expectations, these female walkers' achievements and rebellious spirit confirm Cresswell's argument that:

> mobility bears a number of meanings that circulate widely in the modern Western world. Mobility as progress, as freedom, ... sit side by side with mobility as ... resistance .... Movement is rarely just movement; it carries with it the burden of meaning and it is this meaning that jumps scales .... Movement is rarely just about getting from A to B. The line that connects them, despite its apparent immateriality, is both meaningful and *laden with power*.
>
> (2006, 1–2; 6; 9; emphasis added)

This chapter therefore seeks to highlight this particular form of gendered exclusion, which mainstream cinema has regularly reflected and perpetuated, especially in certain generic contexts, such as the adventure film or the road movie. As Brian Taves (1993) demonstrated in his study *The Romance of Adventure*, the natural outdoors stands, by definition, as the cinematic space for action, adventure, and exploration. This is, without a doubt, a setting

DOI: 10.4324/9781003468028-6

from which women have traditionally been barred. While nature is usually coded as feminine (and, at the same time, women are constructed as existing closer to [Mother] nature than men due to their child-bearing capacity), the wild is also, somewhat paradoxically, not considered to be a space where 'civilized' women should be present (at least on their own) as it is fraught with all sorts of dangers, from precipices and wild creatures to, certainly, (male) human predators.[1] It is a space fit for the testing of true manhood, particularly white manhood, not womanhood.[2] Women belong in domesticated, bounded natural spaces such as the garden or the park, that is, those spaces where nature is kept under control (and where women can also be kept under control). In broad ideological terms, and as much feminist scholarship has contended, wild nature represents an untamed space that needs to be subdued and conquered solely by men (Merchant 1996; Alaimo 2000; Plumwood 2003; Evans 2002; Puleo 2022). For this reason, even though female adventurers, explorers, and walkers do and did exist, they have been culturally marginalized in favor of their better-known male counterparts, hence the parallel lack of adventure narratives headed by female characters (Robinson 2001; Pryce 2015; Thornham 2019; Guignard 2019, 4). This lack of representation signals that cultural constructions of the wild "serve to empower some members of our society while simultaneously disempowering others," which means that "ideas of nature ... are never neutral" (Evans 2002, 181).[3]

In her aforementioned study, Ceuterick affirms that certain films, such as the ones I propose to analyze here, can produce "spatial imaginaries ... within which processes of change may occur" as the women protagonists decide to 'trespass on' and 'occupy' a hostile environment in which they supposedly do not belong (2020, 5). The protagonists in *Wild* and *Tracks* are drawn to wild natural spaces because they are, to quote Alaimo, "undomesticated" in the sense that they choose to stand apart from the bounds of the domestic sphere that often constrains women's agency and movement; these spaces are untamed and thus serve as apt terrain for "female insurgency" (2000, 16). As Alaimo concludes, "[d]espite its role as the bedrock of oppressive ideologies, nature has also been a space for feminist possibility, an always saturated but somehow undomesticated ground" (23).

Thus, I aim to vindicate wild, natural outdoor spaces as sites for female adventure, survival and, crucially, entitlement and emancipation, though not necessarily for female solidarity, as the women in the films that I refer to willingly undertake their hikes on their own. Despite the initial patriarchal skepticism, even hostility, that is invariably present in these narratives (and in society at large) regarding women's capacity to cope in the wild (Glotfelty 1996; Heddon and Turner 2012; Warren 2015; Guignard 2019), these women choose to take a double risk as what is potentially a dangerous activity in itself is presented as even riskier for women walking alone, which increases the protagonists' self-affirmation when all adversity is successfully overcome. In this sense, I argue that these films engage with (eco-)feminism[4] as the wilderness is constructed as a space that women can and should occupy so that changes in the gendered nature of certain spaces (the wilderness itself) and practices (mobility) may occur.

Finally, I propose to analyze this female empowerment process by also referring to the cultural meanings that walking and hiking have accumulated over the centuries (Amato 2004; Nicholson 2011; Gros 2014; Solnit 2014) without losing sight of contemporary ideas regarding biophilia, i.e. the human love of nature (Kellert and Wilson 1993) and nature therapy as an effective route to healing, self-affirmation, and liberation (Lorimer 2011, 23–24; Williams 2017; Jones 2021). These female protagonists are not afraid to listen to the 'call of the wild' and follow their biophilic impulses, which modern anthropocentric societies have so successfully repressed. As a result, not only do they manage to rebuild their identity, but they also find themselves a home in the wild. Like John Muir, they too find that going out is really going in.

Walking and Trekking as Gendered Practices

In the introduction to his work *On the Move*, Tim Cresswell (2006) provides a historical interpretive framework for mobility as an essential human practice. As a whole, he does not apply a gender perspective to his analysis and therefore refers to *people*'s mobility in different historical periods as if it applied to men and women equally.[5] Still, there are fleeting moments when he acknowledges the profoundly gendered nature of mobility. For example, when he refers to the rise of tourism in the West, Cresswell mentions the popularity of the *grand tour* among privileged *men*, and concludes that those early tourists depended "on the relative immobility of those who service the new leisure class—those who are stuck in the picturesque European south as well as those (mostly *women* in the early years) who were left at home" (2006, 15; emphasis added). Although Cresswell does not delve into the matter much further, the obvious implication is that women (and other social groups different from privileged white heterosexual men) accessed the right to mobility (and so many others) only belatedly. As mentioned above, mobility, as a cultural practice, is about power and inequality, yet the male perspective is often naturalized as the default position, which neutralizes the gendered nature of mobility. In the remainder of this section, I shall be reviewing some important texts that have dealt with walking in philosophical and cultural (though not necessarily feminist) terms. Furthermore, I will be referring to the meanings that trekking, as a distinct form of walking, has accrued in Western culture.

As is well known, H. D. Thoreau[6] spent some time living in a hut in the forest, close to his family home in Concord, Massachusetts. From this hut he used to undertake long walks, and the thoughts that these rambles (and the many other walks he took during his life) inspired in him are gathered in his

1851 essay *Walking*. At its very beginning, Thoreau accepts that walking in the wild, which he regards as his occupation, is an activity only available to (certain) men:

> I wish to speak a word for Nature, for absolute freedom and wildness, as contrasted with a freedom and culture merely civil .... If you are ready to leave father and mother, and brother and sister, and wife and child and friends, and never see them again ... then you are ready for a walk .... No wealth can buy the requisite leisure, freedom and independence which are the capital of this profession.
>
> (2017 [1851], 26–27)

Thoreau goes on to explain how sorry he feels for the "mechanics and shopkeepers" who have to spend all their time at their places of work: "I think that they deserve some credit for not having all committed suicide long ago" (28). He concludes the introduction to his essay by showing some sympathy for women: "How womankind, who are confined to the house even more than men, stand it I do not know; but I have ground to suspect that most of them do not *stand* it at all" (28; emphasis in the original).

Stand it or not, most women in the West have been confined to the safety of domestic spaces, whether the home or the garden, until relatively recently. Contact with the wilderness, as Thoreau admonishes, "will no doubt produce a certain roughness of character" whereas staying at home will "produce a softness and smoothness, not to say thinness of skin" (29). This long-standing cultural expectation has recurrently led to restrictions being imposed on women's mobility, freedom, and agency. Still, for Thoreau, one cannot "live deep and suck out all the marrow of life" (2004 [1854], 91), as he put it in *Walden*, until one experiences the wilderness and is enmeshed in it *first-hand*. As he exclaims: "How near to good is what is wild! ... Hope and the future ... are not in lawns and cultivated fields ... but in the impervious and quaking swamps" (2017 [1851], 39). Yet, the sort of free, uncomplicated access to nature that Thoreau philosophized about remained a 'wild' possibility for most women until recent times.[7] Women internalized and imposed such restrictions on themselves so as to avoid being labeled as 'wild women' or 'witches' and it was only much later that some 'thick-skinned' women started to challenge prevailing norms by claiming natural spaces as their own territory to tread.

Thoreau was explicit as regards the exclusion of women from wild spaces, but this is rare. When one takes a cursory look at some recent literature on walking as a cultural practice it is easy to appreciate a 'default' male perspective. Apart from important exceptions such as Solnit (2014) and Guignard (2019), both female authors, the monographs I have had access to deal with walking as a gender-*neutral* practice. That is, (male) authors tend to express their insights and experiences as men as if they applied to all members of the human race equally. Here are some examples:

> When I find myself in a new place I explore it on foot .... Maybe it's a way of *marking territory*, of beating the bounds—setting foot in a street makes it yours in a way that driving down it never does .... If you live in a big city ... walking is often preferable to public transport. You don't have to wait for trains and buses that may not arrive .... *When you walk you are your own boss.*
>
> (Nicholson 2011, 17–18; emphasis added)

> It is impossible to be alone when walking, with so many things under *our gaze* .... All those fields, houses, forests, paths, *all ours*, for us. We have *mastered* all that by our ascent .... Who could feel alone when *he possesses* the world? Seeing, dominating, looking mean possessing.
>
> (Gros 2014, 55; emphasis added)

While I do not intend to speak for all women, I believe it would be rather challenging to find one who actually feels like they are their "own boss" when they walk alone in a place they do not know, especially at night. Meanwhile, the second extract can safely be considered yet another manifestation of Mulvey's male gaze theory. These examples of male bias demonstrate that, even in this day and age, female mobility cannot be taken for granted.

In the quotation above, Gros is specifically referring to trekking and mountaineering. These outdoor activities became popular in the Romantic period, during which the reverence for nature became an integral part of the cultural landscape.[8] As Solnit (2014, 134) explains, this particular form of walking became popular among privileged men and indeed many climbing clubs did not accept women initially, the Sierra Club in California, founded by John Muir, providing a rare exception (153). Yet again, women accessed the risky world of mountaineering only later, and, as is to be expected, encountered much social (particularly male, but also female) prejudice.[9] This kind of bias, which still endures today,[10] slowly but surely became eroded thanks to intrepid women, unafraid of stepping outside, and those who followed in their footsteps. Guignard believes that women's comparatively smaller presence in outdoor recreation activities, whether mountaineering, trekking, or just camping, is once again due to the limitations placed on their freedom of movement: "mobility has never been a gift freely given to women, and a tent is the ultimate symbol of independence: a house she can take anywhere and provide for herself" (2019, 23). As she concludes, outdoor activities need not be the answer to every woman's woes, but at least they all should feel that they are options that every woman can pursue (29).

This is precisely the route that the protagonists in *Wild* and *Tracks* decide to take. In what follows, I offer a reading of these films, which I have framed not only within ecofeminism and the theories of mobility/walking that I have sketched above, but also within the bounds of film genre theory.

Women and Adventure Narratives

When considering films dealing with adventure (and survival) in the wild, examples like *Deliverance* (United States, 1972), *Jaws* (United States, 1975), *The Edge* (United States, 1997), *Castaway* (United States, 2000), *Into the Wild* (United States, 2007), *The Way Back* (United Kingdom/United States, 2010), *Life of Pi* (United States, 2012), and *The Revenant* (United States, 2015) come readily to mind. Upon closer consideration of these titles, there is an element that becomes evident, and that is the lack of women playing the main roles. Indeed, a quick internet search for adventure films reveals the intensely masculinist outlook of this type of film. Although it tends to overlap with various other genres, the survival-in-the-wilderness film is generally thought of as a subgenre of the action-adventure narrative, from which female leads have remained largely, though not entirely, absent (Ingram 2004, 36). There may be several reasons for this. As already mentioned, femininity does not seem to be culturally compatible with such concepts as mobility, adventure, survival, and, more generally, life in the wilderness. In patriarchal western societies, femininity has been linked to domesticity while at the same time Nature has been socially constructed as a resource to be tamed and exploited, which has provided fertile ground for the representation of myriad narratives of mastering masculinity. In addition, adventure narratives have sometimes (via the war film or the safari film, to name but a few) foregrounded the motif of dominance, conquering, and imperialism, an endeavor that has been regularly reserved for white men (Taves 1993, 122).[11]

However, what happens when these kinds of narratives are led by women? What are the implications in terms of gendered representations? In order to answer these questions, I aim to analyze the motives that drive the women protagonists into the wild, together with the ways in which they adapt to and cope with their environment. Though normally absent from the aforementioned lists of 'essential' adventure/survival films, several titles are worth mentioning: *Walkabout* (Australia, 1971), *Rabbit-Proof Fence* (Australia, 2002), *Meek's Cutoff* (United States, 2010), *The Hunger Games* (United States, 2012), *Tracks* (United States, 2013), *Wild* (United States, 2014), *Endless Night* (Spain/France/Bulgaria, 2015), *Mad Max: Fury Road* (Australia/United States, 2015), *The Shallows* (United States, 2016), and *Edie* (United Kingdom, 2017) are among them. It is in the last twenty years or so that more and more adventure narratives led by resourceful women have been produced. Nevertheless, female leads, though a conspicuous minority, have existed in adventure narratives since the 1930s at least. Indeed, as Taves (1993, 135) asserts in his account of the genre, there is nothing in its narrative structure that precludes the existence of women leads, so the explanation for this absence needs to be found elsewhere.

The aforementioned films exhibit traits of the classical adventure narrative while also borrowing elements from other generic contexts, the Western, science fiction, horror, and the road movie, among them. All of them showcase various ways in which competent women step into and engage with nature and its challenges. Largely speaking, these narratives present the wild in ambivalent terms. While it is understood to pose numerous dangers for the women protagonists, from broken toenails to dehydration, to rape and death, direct contact with nature is at the same time considered to be therapeutic, liberating, and necessary. In this particular context, *Tracks* and *Wild* stand out.

In these trekking adventure narratives there is a strong focus on the idea of the journey. Though the protagonists of the films resolve to go on solo treks for different reasons, these narratives focus heavily on such Romantic motifs as individuality, independence, mobility, freedom, the embrace of simple pleasures, such as the beauty provided by the contemplation of open spaces or the desire to break away from one's previous social environment and re-connect with nature in order to achieve personal renewal. These are the very traits that characterize the, yet again, thoroughly masculinized road movie genre (Cohan and Hark 1997, 3). As Brereton observes in connection with the genre,

> sublime landscape is foregrounded in road movies in order to promote deep ecological expressions of oneness with nature .... Landscape [is] a space of pilgrimage which romantics ... believed helped to reveal clearer moral truths concerning human nature.[12] Only when nature is fully accepted and appreciated and the protagonists become part of nature can they finally meet their destiny.
>
> (2004, 91 and 103)

I believe these Romantic aspects of the road movie are also present in both of the films under analysis here since their adventurous protagonists embark on long and risky treks with roughly similar purposes. To use the words of Cheryl Strayed (Reese Witherspoon) in *Wild*, they voluntarily choose to place themselves "in the way of beauty" so as to redefine their existence and their relationship with the natural outdoors in their own terms. In adventure narratives led by men, such as the ones mentioned above, this is rarely the case. On the other hand, in the road movie genre, the journey, rather than the final destination, represents the central motif or narrative concern. Thus, if the road movie constitutes a type of *bildungsroman* that focuses on personal growth resulting from embarking on a journey, then films about trekking like *Wild* and *Tracks* can be said to be closely related to the road movie genre, albeit with the Pacific Crest Trail in the US and the deserts of Western Australia substituting for the open road and walking becoming the preferred 'back-to-basics' means of transportation.

Natural landscapes are no doubt important in the films under analysis here, no matter how harsh or hostile they may turn out to be. However, unlike in other examples of the adventure genre, here the land is not represented so as to convey the idea of mastery and possession, which Gros (2014) blatantly referred to above. In *Wild* and *Tracks* the landscape is not represented as a site to be subdued but as a refuge, a sacred place even, to go to alone and respectfully connect with in order to rebuild one's identity by affirming and accepting oneself. At some point, Robyn Davidson (Mia Wasikowska) in *Tracks* explains, "the trip wasn't meant as an adventure in the sense of something being ... conquered." And in fact, these heroines are never entirely in control of their surroundings: the threat of dehydration, exhaustion, rape, or death is ever present. Instead, the landscape is more Romantically represented as a site in which to submerge yourself in order to become liberated (from your own demons, from patriarchal expectations, from social prejudice, from the company of others) and start again in your own terms. By doing so, these female characters manage to leave their baggage behind, reshape their identities, and develop a deeper connection with the wild, which allows them to accept their existence as the plain members, not conquerors, of the Land that Aldo Leopold proposed human beings should become (1979, 204).

In any case, as already explained, adventure narratives, whether real or fictional, have not usually incorporated female protagonists. Like Guignard (2019) above, author and adventurer Lois Pryce laments that real-life female adventurers like herself have generally been ignored. This is because, even today, "there is something discomfiting about the image of a ... capable female.... A wild, free woman is a dangerous concept" (2015). As Pryce suggests, this implies that women's adventure narratives tend to perpetuate the idea that females need a potent reason, a real motive, to hit the trail. Unlike their intrepid male counterparts, women cannot *possibly* be doing it out of curiosity, to test their abilities or just for the heck of it. In other words, they constantly need to justify their presence in and their desire to explore wild, open spaces.

Pryce also believes that this general lack of interest in female adventurers derives from the fact that "the reality of life on the road is too primal for public consumption—it's messy and crude ... and occasionally hairy, in both senses of the word, all of which are regarded as unfeminine and therefore, unpalatable for mainstream cinema," hence the comparably higher popularity, especially among male spectators, of the hypersexualized 'action-babe' formula, the 'post-feminist' action-adventure subgenre *par excellence* (O'Day 2015). *Wild* and *Tracks* choose to emphasize their protagonists' messiness and flaws instead, thereby refusing to present them as objects for contemplation. Thus, by placing their resolute, though imperfect, women protagonists at the very center of the narrative universe, these films succeed in opening new spaces for the representation of female subjectivity and agency in contemporary film.

Wild

It is also fitting to read *Wild* in the light of Hilary Radner's (2017, 1–6) conceptualization of the new woman's film,[13] i.e. a "little picture" originating from a studio specializing in independent and foreign productions headed by a prestigious female star of the "post-feminist" era who is willing to participate in small projects for a reduced fee. Thus, it is probably no coincidence that Reese Witherspoon, one of the most important stars in the so-called post-feminist era (Rowe Karlyn 2011, 127–160),[14] chose to produce and star in *Wild*. Much like Cheryl Strayed, author of the book on which the film is based, Witherspoon was looking for ways to redefine herself as a performing artist so as to leave her 'girly' baggage behind. She hoped to do so by playing against type as a frequently unpleasant, working-class sex and drug addict going through divorce and bereavement.[15] The foregrounding of consumerism and female sexuality that often characterizes the contemporary woman's film is certainly taken to the extreme in this movie. And it is in fact Sheryl's sex and drug addictions that make her reach rock bottom. In a moment of desperation when she realizes she is pregnant, a book about the PCT (Pacific Crest Trail) in a shop triggers her biophilic impulse and inspires her to take the 3,000-mile trek. She resolves to embark on a sort of pilgrimage in order to find atonement in nature and become "the woman [her] mother thought [she] was."

Wild therefore provides a good example of the mainstream female-led narratives that Pryce criticizes above, as it is as a result of inner turmoil that Sheryl retreats to the wilderness and chooses to put herself "in the way of beauty." This is her 'justification' for being present in the public space of the PCT. The fantastic, sublime landscape of the Trail becomes the therapeutic antidote Sheryl's troubled psyche needed. As Simon and Garfunkel sing in "El Cóndor Pasa (If I Could)," which features prominently in the film,[16] Sheryl "would rather feel the Earth beneath [her] feet" than continue spiraling downward. She thus follows in the footsteps of the Romantics, whose longing for direct contact with nature (or biophilia) as an antidote for social ills and human anomie has long been researched by scientists (Kellert and Wilson 1993; Williams 2017; Jones 2021). It is thanks to her incursion into the wild and her forging a sense of belonging with and in the land that Sheryl is able to come to terms with her mother's untimely death, thus achieving self-knowledge and, more importantly, self-control and self-acceptance.

Still, the path to atonement is anything but easy. The PCT is hard enough for regular hikers, let alone for ordinary women like Sheryl. The film is full of mildly comic moments that confirm that she is completely out of place in the wilderness. Her walking boots are too small, which leads to her lose some of her toenails; she brings the wrong kind of stove fuel so she has to eat "cold mush" for quite a few days; she is unsure about how to pitch a tent; last but not least, her backpack is huge as it is filled with all sorts of unessential items and she needs the help of a man who is actually *not* a regular hiker to sort this out for her. Certainly, all of these small 'catastrophes' point at women's

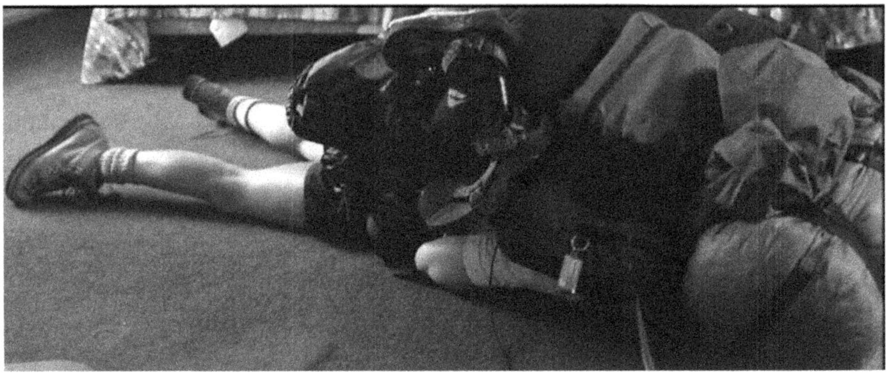

Figure 5.1 Sheryl's initial outdoor incompetence in action.

regular absence from and purported inadequacy for outdoor activities, which has been accounted for in previous sections. The first sequence, during which Sheryl pants her way to a mountain pass, has to pull her painful toenail, and loses her walking boot, establishes her incompetence in the wild from the get-go.

At the same time, however, the suffering she undergoes during the trek places her (mis)adventure in the context of mountaineering as religious experience/pilgrimage, which has been referred to above. The fact that, unlike other fellow, more experienced male hikers, she completes the trek, i.e. she fulfills her promise, eventually vindicates her competence and, most importantly, her 'right to roam' in the wilderness.[17]

Yet, this is not done without difficulty. Apart from dealing with all her flaws as a hiker, Cheryl has to face certain situations that generally put women off participating in outdoor adventure activities, however rare. The day before she starts the trek she checks into a motel and the receptionist is skeptical about her desire (and certainly her capacity) to walk alone all the way; she takes it for granted that a male companion will eventually join her, despite Sheryl's rebuff. In the background, news about the O.J. Simpson trial is overheard, which is a reminder of the fact that violence against women is a possibility always lurking. And indeed, Sheryl has to deal with this threat herself several times, with different levels of intensity. Firstly, when hungry, cold, and thirsty she asks a farmer to give her a lift to a nearby restaurant so that she can get a hot meal. Even though he turns out to be harmless, during a tense conversation full of sexual undertones ("What kind of woman are you? ... A wild woman?") she feels the need to make up a story about her husband being only a few miles ahead; secondly, when a refuge warden constantly refers to her as "good-looking" and agrees to help her only if she has a drink with him; and most importantly, when she narrowly escapes getting raped after a dangerous encounter with two hunters.

Despite these negative experiences, Sheryl does not give up and instead focuses on her objective, that is, "walk[ing] unafraid"[18] in order to test herself, find her place in the wild, and honor her mom. A self-professed feminist, Sheryl reads her way along the PCT (Adrienne Rich's poem "Power" is only one of her many inspirations) and reminisces about studying feminist writers with her recently deceased mother, a victim of gender violence who was strong enough to leave her abuser. She found solace in nature and horse therapy, which is what prompted Sheryl to undertake the PCT trek. Despite the initial hardships, she gradually starts to find her space in the wilderness: "I'm lonelier in my real life than out here," she confesses to another rare solo female hiker as they enjoy the sunset. When Sheryl asks her why she is doing the trek she says: "I need to find something in myself. I think the trail is good for that. I mean, look, this has the power to fill you up again if you let it," which sums up the rationale behind nature therapy nicely. The sunset reminds Sheryl of her mom, a trauma survivor who believed that despite life's hardships "there is a sunrise and a sunset every day" and one must always strive to place oneself "in the way of beauty."

When she is about to complete the trek, Sheryl calls her estranged brother to ask him to "walk a little" as she can sense that this hard but worthwhile experience is having the desired effect. While in the initial stages of the trek Sheryl felt clearly out of place in the wild, in the final ones she is clearly at home and at peace; she is able to sleep rough and is happy to find other fellow creatures covering her face and sleeping bag in the morning. Self-forgiveness and self-acceptance are also what characterize her final speech: "What if I forgive myself? ... What if all those things [addictions] are what got me here? What if I was never redeemed? What if I already was?" By the end of her walk, Sheryl has become a competent, resilient, and tougher hiker who has started to overcome trauma. Sheryl's adventure has allowed her to rewrite her story in her own terms, thereby opening a space for feminist

Figure 5.2 "There is a sunrise and a sunset every day and you can choose to be there for it."

possibility. The spectator also learns that some years later she became married once again and had two children. Thus, while she has shown tremendous courage along the way, her decision to embrace marriage and motherhood also brings us back to more conventional territory, which confirms Radner's aforementioned appreciation regarding the ultimately conservative tendencies of the contemporary woman's film. As portrayed in this ambivalent Hollywood production, it seems that the biophilic impulse has not entirely managed to trump the post-feminist one.

Tracks

In contrast, Robyn Davidson, in *Tracks*, considers herself a perennial nomad who can't be at home anywhere. Importantly, this narrative does not clearly force its protagonist to provide a potent reason for embarking on a 2,000-mile trek across the Australian outback. Like *Wild*, the film is based on a true story which went on to become a bestseller and although there are occasional flashbacks that point to a possible source of childhood trauma in the protagonist's life, *Tracks* does not elaborate on this issue as extensively as the more melodramatic *Wild* does. Instead, when Robyn explains why she wants to hit the trail she says, "why not?"[19] At some point, she also explains "I just want to be by myself."

She mentions her father, who trekked across the Kalahari in his youth and "was happiest on his own out in the bush," as well as Australian explorer Ludwig Leichhardt, who disappeared when trying to cross Australia on horseback, as her real sources of inspiration. In essence, then, the protagonist's reasons seem to emanate from her pure desire to experience raw nature alone and become an independent woman, free from the conventions that her society imposes on her. The book on which the film is based was published during the 1970s, that is to say, the time that most represents the second wave of the feminist movement. In this way, *Tracks* draws inspiration from the 1970s feminist 'independent woman's film' rather than the contemporary woman's film (Radner, 2017), which is the case for *Wild*. Robyn's refusal to accept National Geographic photographer Rick Smolan (Adam Driver) as a trekking companion, let alone a romantic one, together with the film's overt denunciation of sexism (and racism) clearly point in this direction.

At the start of the film, Robyn hints at her reasons for embarking on her trek:

> I'd always been drawn to the purity of the desert; its hot wind and wide-open spaces ... but mainly I was bored with life in the city, with its repetitions, ... my half-hearted attempts at jobs and various studies and I was sick of carrying around the self-indulgent negativity that was so much the malaise of my generation, my sex, and my class.

Her desire to defy social expectations and rid herself of such disenchantment, but also her biophilic impulse, prompt her to act and retreat to the desert:

"I believe that when you've been stuck in one point too long it's best to throw a grenade where you're standing and jump … and pray." These words are uttered during a sequence in which we see Robyn tending bar, i.e. a *public* house. There is a conspicuous lack of women in this homosocial space, except for an Aboriginal woman who is beaten and kicked out of the pub, to Robyn's dismay. This sequence suggests that even though *white* women were only allowed in such public spaces as bars so as to serve men, Aboriginal women were much more violently excluded from the social arena. One therefore misses reference to her 'race,' as well as her sex and class, in her denunciation of generational self-indulgency above. That said, even though Robyn is eventually able to take advantage of the privilege that her 'race,' if not her sex, affords her in order to embark on her adventure, she eventually has to rely on Aboriginal elder Mr. Edie's (Rolley Mintuma) knowledge and escorting in sacred sites in the desert,[20] as well as Rick dropping water cans along the way,[21] apart from her camels and dog, to be able to realize her dream.

As a white woman, then, Robyn mobilizes her racial privilege to earn money to finance her dream, but also extreme, expedition. She tends bars and works on a camel farm, where she is exploited and conned by a particularly mean, misogynistic boss, who describes her plan as "ridiculous." She ends up working for another camel farmer who, despite his initial disbelief ("you don't have to be unlucky to die out there," he says), treats her respectfully and fairly. When she obtains the camels that she needs, she applies for sponsorship from National Geographic, who accept on condition that one of their reporters can photograph the expedition, which she only agrees to grudgingly as it is the extreme solitude of the desert that she craves.

During the expedition she has none of the dangerous encounters that Sheryl Strayed had to face and in fact all the men she has contact with are helpful and kind. However, she becomes a reluctant international celebrity, "the camel lady," and the attention she receives overwhelms her, threatening her mental stability at times. This, together with the utter harshness of the trek itself and the traumatic death of her dog companion Diggety (Special Agent Gibbs) put the expedition in jeopardy, but only briefly. Like Sheryl's, her resilience is astonishing, and she manages to complete the trek, even if with a little help from her human and nonhuman friends. Even though she needs to be escorted temporarily by Mr. Edie through the sacred sites she needs to traverse, her ability to complete the expedition successfully establishes, like in *Wild*, her competence. She rightfully earns 'her right of way' in the desert, but also develops a strong sense of belonging in this harsh but magnificent natural space. The film's cinematography emphasizes this through its frequent use of shots in which she seems to blend in with the landscape.

Much like the camels imported by European settlers that have adapted to and now roam freely in the Australian outback, Robyn 'goes feral' and chooses to leave her previous, conventional life behind and find her place in the desert. By claiming her right to go 'walkabout,' she reshapes her female identity in this natural space, where she finds refuge and comfort. Despite the harshness of the experience, there is no doubt that Robyn has also found herself a home.

Figure 5.3 The harshness of the trek, but also Robyn's right to inhabit the wilderness, is inscribed on her burnt, drained body.

Conclusion

In this chapter, I have tried to analyze two recent, refreshing examples of the adventure genre from an ecofeminist perspective. Despite their differences, these films display a common interest in individual female self-realization through close contact with nature. Sheryl and Robyn dare to follow their biophilic impulse and find peace and comfort in their 'pilgrimage' but also in their isolation though, while Sheryl overcomes trauma and rebuilds her life in more conventional terms, Robyn's lot is unclear. Her physical destination has been reached but her emotional quest has not ended. Thus, in her case, we can conclude, as the film does, that "camel trips do not begin and end, they merely change form." Despite the different paths that the protagonists eventually choose to walk, ultimately reflecting different moments in the history of women's struggle, both films share a feminist concern for the affirmation of female identity in the wilderness.

By following their biophilic impulses, claiming and finding their space in the wild, the protagonists of these two films are able to achieve self-knowledge and overcome their respective crises. To paraphrase John Muir at the beginning of this chapter once again, by going out these two women are actually going in, that is, going home. Their drive to experience nature first-hand allows them to break the mold and experience the restorative, liberating powers of nature from which many women have been culturally and forcefully excluded. In these films, the wilderness is constructed as the 'haven in the heartless world' that modern, materialistic societies have become in the era of the Anthropocene. Importantly, then, these two narratives provide grounds for hope in the ability of women to feel safe and at home in the natural outdoors, but more generally in the human capacity to appreciate nature not only in utilitarian terms but also as the vital refuge we should all regularly retreat to and work to protect.

Finally, films like *Tracks* and *Wild* lead the adventure genre in a new direction. The relative success of these 'trailblazing' films, both among critics and at the box office, indicates that this new territory deserves further exploration. Hopefully, it will be charted in the years to come as more and more willful female leads claim their right to 'take back the walk.'

Funding

This chapter is one of the results of research project "Crisis and Hope: Structures of Feelings and their Representation in Contemporary Cinema," PID2020-114338GB-I00, funded by MICIU/AEI/10.13039/501100011033.

Notes

1 Still, statistically speaking, the outdoors is *not* a more dangerous space for women to tread. In connection with this, and with US women's perceived lack of interest in outdoor activities, Guignard stresses: "walking in the wilds, women ... carry the warnings and threats of society in our heads—that by venturing out into wild spaces we are asking for it. That we are possibly more sexualized and objectified in such spaces, even as we attempt to escape such cultural labels and restraints ... Women have internalized this spatial patriarchy, voluntarily limiting their mobility without conscious consideration of which spaces are really the most dangerous .... The trail far from the parking lot is statistically less dangerous than a woman's home" (2019, 4 and 24).

2 As Guignard reminds us, the first national parks in the US (and in the world) were in large measure created so as to facilitate the preservation of "real" US manhood. President Roosevelt, and other significant figures such as Frederick Jackson Turner, feared that the closing of the frontier would result in US males becoming "overcivilized, [i.e.] feminized and Europeanized." He also believed that separating older boys from their mothers' influence so that they could "test themselves against the wilderness was the only way to make them real men" (Guignard 2019, 80-81).

3 Guignard (2019, 29) laments that those rare media accounts of female achievements in the masculinized outdoors construct these women as "outliers, freaks of nature," that is not your average Jane, which can deter other women who have not had regular access to nature from following in their footsteps.

4 As Plumwood (2004), Gaard (2010) and Puleo (2022) explain, ecofeminism is diverse as a discipline although there is a general tendency to focus on the clear parallelism that can be established between the subjugation of women and the exploitation of nature. While this analogy is potent, it does not necessarily signify that women are closer to nature. This is a patriarchal construction that feminism has fought to dismantle for a long time. Unfortunately, this struggle has managed to devalue the idea of nature even further: "if woman's perceived proximity to nature is responsible for her oppression, then her liberation, it would seem, is contingent on her distance from nature" (Alaimo 2000, 3 and 44).

5 While my argument specifically focuses on the exclusion of women from the literature on walking, it could also be applied to the absence of other non-privileged groups, such as those that are non-white, non-heterosexual, non-binary or those with disabilities, among others.

6 According to Amato, Thoreau has become "the patron saint of localists, ecologists, and country walkers" (2004, 151).

7 As Solnit explains: "[l]egal measures, social mores subscribed to by both men and women, the threat implicit in sexual harassment and rape itself have all limited women's ability to walk where and when they wished .... If walking is a primary

cultural act and a crucial way of being in the world, those who have been unable to walk out as far as their feet would take them have been denied not merely exercise or recreation but a vast portion of their humanity" (2014, 234; 240; 245).

8 English Romantic poet William Wordsworth is credited with inaugurating the practice of walking in the wilderness for pleasure, rather than as a necessity (Solnit 2014, 82; Gros 2014, 209). As Amato states, his poetry "dissolved walking's perennial association with necessity, poverty and vagrancy .... [R]omantically inspired walkers took to the road by choice ... to the bewilderment and disdain of those who walked out of necessity" (2004, 106 and 107).

9 By way of an example, Guignard (2019) mentions the case of pioneer Miriam O'Brian, the first woman to climb the Grépon mountain in the French Alps *manless*. The assumption at the time was that "if a man was around, the woman did not do it all on her own" (2019, 130). But she did, with her climbing partner Alice Damesme. After completing this amazing achievement some male voices lamented that "the Grépon has disappeared ... Now that it has been done by two women alone, no self-respecting man can undertake it" (132).

10 At the time of writing, a group of 200 female rock climbers have gathered in a village near my hometown to celebrate and vindicate women's access to this sport.

11 Amato (2004, 118) and Solnit (2014, 138) have established links between mountaineering, which as mentioned above was initially an activity dominated by privileged white men, and the ethos of militarism and imperialism.

12 Amato (2004, 117-118) also explains that ever since the Romantic period, mountains have provided some hikers with feelings of transcendence and a sense of closeness to divinity, thereby turning them into "dark-green" religious pilgrimage sites.

13 Radner argues that a sustained focus on female subjectivity is the defining characteristic of the "post-feminist" spate of films that she labels the "new woman's film." These low-key films tend to "address the ... shifting terrains of contemporary sexual expectations ... while remaining essentially conservative in [their] notions of human relations and [their] emphasis on family and parenthood" (2017, 3).

14 Radner (2017, 18) considers her to be an essential "girly girl."

15 To a certain extent, this persona was retrieved in the series *The Morning Show* (2019-ongoing), Witherspoon's current streaming platform success.

16 *Wild* certainly follows the 'amazing music against amazing background' convention that is typical of the US road movie.

17 In connection with Witherspoon's resilient star persona, Rowe argues that she epitomizes the "Third Wave star in her refusal to appear a victim .... Bad things happen to the characters she plays, and they make mistakes, but they come out on top" (2011, 133).

18 *Walk Unafraid* is the REM song that closes the film's credits.

19 This is reminiscent of a well-known anecdote in mountaineering history. When George Mallory was asked for his reasons to climb Mount Everest he purportedly replied "because it's there" (Pryce 2015).

20 Reluctantly, but respectfully, Robyn accepts Mr Eddie as a trekking companion in order to avoid a long detour. Indeed, the fact that they cannot initially communicate verbally helps Robyn to adapt to the situation. Eventually, however, she becomes fond of the man who introduces her to Aboriginal wisdom about the ways of the land. It is thanks to Eddie that Robyn can survive the last stretch of the trek on her own, which involves walking for two months without a human settlement in sight.

21 Interestingly, in Davidson's autobiography it is not Smolan but Glendle Schrader, a friend she meets during her trek and also appears in the film, that provides the water. This important narrative shift suggests that the director and scriptwriter

thought it was necessary to reinforce Robyn's dependence on her would-be lover rather than a friend in order, perhaps, to make her a more sympathetic or 'palatable' character. The film main publicity poster also hints at a love relationship that is not actually realized.

Works Cited

Alaimo, Stacy. 2000. *Undomesticated Ground. Recasting Nature as Feminist Space.* Ithaca and London: Cornell University Press.

Amato, Joseph. 2004. *On Foot: A History of Walking.* New York: New York University Press.

Brereton, Pat. 2004. *Hollywood Utopia. Ecology in Contemporary American Cinema.* London: Intellect Books.

Ceuterik, Maud. 2020. *Affirmative Aesthetics and Wilful Women. Gender, Space and Mobility in Contemporary Cinema.* Cham, Switzerland: Palgrave MacMillan.

Cresswell, Tim. 2006. *On the Move. Mobility in the Modern Western World.* New York and London: Routledge.

Cohan, Steven, and Ina Rae Hark. 1997. "Introduction." In *The Road Movie Book*, edited by Steven Cohan and Ina Rae Hark, 1–14. London and New York: Routledge.

Evans, Mei Mei. 2002. "'Nature' and Environmental Justice." In *The Environmental Justice Reader. Politics, Poetics and Pedagogy*, edited by Joni Adamson, Mei Mei Evans and Rachel Stein, 181–193. Tucson: University of Arizona Press.

Gaard, Greta. 2010. "New Directions for Ecofeminism: Toward a More Feminist Ecocriticism." *Interdisciplinary Studies in Literature and the Environment* 17 (4): 643–665.

Glotfelty, Cheryll. 1996. "Femininity in the wilderness: Reading gender in women's guides to backpacking." *Women's Studies: An interdisciplinary Journal* 25 (5): 439–456.

Gros, Frédéric. 2014 [2009]. *A Philosophy of Walking.* Translated by John Howe. London and New York: Verso.

Guignard, Lilace Mellin. 2019. *When Everything Beyond the Walls is Wild. Being a Woman in Outdoors America.* College Station: Texas A&M University Press.

Heddon, Deirdre, and Cathy Turner. 2012. "Walking women: shifting the tales and scales of mobility." *Contemporary Theatre Review* 22 (2): 224–236.

Ingram, David. 2004 [2000]. *Green Screen. Environmentalism and Hollywood Cinema.* Exeter, Devon: University of Exeter Press.

Jones, Lucy. 2021 [2020]. *Losing Eden. Why Our Minds Need the Wild.* Dublin: Penguin Books.

Kellert, Stephen R., and Edward O. Wilson, eds. 1993. *The Biophilia Hypothesis.* Washington, D.C. and Covelo, California: Island Press.

Leopold, Aldo. 1979 [1949]. *A Sand County Almanac: With Other Essays on Conservation from Round River.* Oxford: Oxford University Press.

Lorimer, Hayden. 2011. "Walking: New Forms and Spaces for Studies of Pedestrianism." In *Geographies of Mobilities: Practices, Spaces, Subjects*, edited by Tim Cresswell and Peter Merriman, 19–30. Farnham and Burlington: Ashgate Publishing.

Merchant, Carolyn. 1996. "Reinventing Eden. Western Culture as a Recovery Narrative." In *Uncommon Ground*, edited by William Cronon, 132–159. New York and London: Norton & Company.

Nicholson, Geoff. 2011. *The Lost Art of Walking.* Chelmsford: Harbour Books.

O'Day, Marc. 2015. "Beauty in Motion: Gender, Spectacle and Action Babe Cinema." In *Action-Adventure Cinema*, edited by Yvonne Tasker, 201–218. London: Routledge.

Plumwood, Val. 2003 [1993]. *Feminism and the Mastery of Nature*. London and New York.
Plumwood, Val. 2004. "Gender, Eco-feminism and the Environment." In *Controversies in Environmental Sociology*, edited by Rob White, 43–60. Cambridge: Cambridge University Press.
Pryce, Lois. 2015. "Reese Witherspoon in 'Wild': An all too rare story of a woman on a solo adventure." *The Independent*, January 11, 2015. https://www.independent.co.uk/travel/activity-adventure/reese-witherspoon-in-wild-an-all-too-rare-story-of-a-woman-on-a-solo-adventure-9971130.html.
Puleo, Alicia. 2022 [2019]. *Claves Ecofeminsitas. Para rebeldes que aman a la Tierra y a los animales*. Madrid: Plaza y Valdés.
Radner, Hilary. 2017. *The New Woman's Film: Femme-centric Movies for Smart Chicks*. London and New York: Routledge.
Robinson, Jane. 2001 [1991]. *Wayward Women. A Guide to Women Travelers*. Oxford: Oxford University Press.
Rowe Karlyn, Kathleen. 2011. *Unruly Girls, Unrepentant Mothers. Redefining Feminism on Screen*. Austin: University of Texas Press.
Solnit, Rebecca. 2014 [2001]. *Wanderlust: A History of Walking*. London: Granta Books.
Taves, Brian. 1993. *The Romance of Adventure: The Genre of Historical Adventure Movies*. Jackson: University Press of Mississippi.
Thornham, Sue. 2019. *Spaces of Women's Cinema: Space, Place and Genre in Contemporary Women's Filmmaking*. London: Bloomsbury Publishing.
Thoreau, Henry D. 2004 [1971]. *Walden*. Princeton and Oxford: Princeton University Press.
Thoreau, Henry D. 2017 [1851]. *Walking*. Los Angeles: Enhanced Media Publishing.
Uteng, Tanu Priya, and Tim Cresswell, eds. 2016 [2008]. *Gendered Mobilities*. London and New York: Routledge.
Warren, Karen. 2015. "Gender in Outdoor Studies." In *Routledge International Handbook of Outdoor Studies*, edited by Barbara Humberstone et al., 360–368. London and New York: Routledge.
Williams, Florence. 2017. *The Nature Fix. Why Nature Makes Us Happier, Healthier, and More Creative*. New York: W.W. Norton & Company.

6 Between Manipulation and Catharsis

Living a Life in the Mediated Anthropocene

Ignacio Bergillos

Introduction

The official poster for the 2022 Cannes Film Festival paid tribute to *The Truman Show* (Peter Weir, United States, 1998). Based on the idea that "art and cinema are where contemplation and the renewal of society unravel," the festival justified the decision by pointing to the way in which Peter Weir's film "urges viewers to not only experience the border between reality and its representation but to ponder the power of fiction, between manipulation and catharsis" (Festival de Cannes 2022). Thierry Frémaux, General Delegate of the festival, added that "we thought this gesture of touching the sky, blue sky, is what we all aspire to, especially now" (Festival de Cannes 2022).

This chapter continues this reflection and delves into the ambivalent meanings that cinema offers us in relation to our socio-natural relations. It approaches Weir's film from a position of interest in knowing the material limits that condition our mediated life and the paradoxes we face once we acknowledge that, like Truman, we live a mediated life in the Anthropocene. What are the foundations of that mediated life? How do we deal with the paradox of being an unknowing protagonist in an era in which our actions affect the environment in which we live? How do we deal with the recognition of the important role of the media in our lives and of our actions in the planetary ecosystems? How does the filmic text help in the process of making meaning of the Anthropocene from the perspective of the person who discovers his role and, at the same time, his powerlessness in shaping his environment? Given that a mecca of cinema like Cannes turns to this film, it seems timely to explore *The Truman Show* from the perspective of ecocinema in order to understand how it represents our socio-natural relations and how the film simultaneously illustrates the cultural influence of media and its material embeddedness in our lives (Rust, Monani, and Cubitt 2013).

This chapter is part of a wider project that focuses on the interrelationship between media, communication, and the Anthropocene. It is framed within environmental media studies, the interdisciplinary approach that interrogates the role of digital media in our lives and its connection to the challenges of (and opportunities for) the Anthropocene. Through this focus,

DOI: 10.4324/9781003468028-7

I acknowledge the importance of screen media and media texts in the way we make sense of our environment.

My central argument is that, to resituate ourselves in the Anthropocene, we need to rethink our relationship with media. Cinema is, of course, part of it, and I will use *The Truman Show* in my reflection on what it means to live a media life in the Anthropocene. As a starting point, this chapter will delve into the concerns and conceptual interrelations of two points of departure: first, that today we are living *in* media rather than *with* media (Deuze 2012), and second, that the Anthropocene is a conceptual framework that we can use to reflect on our socio-natural relations in a moment of environmental crisis (Arias-Maldonado 2015). Both premises are based on the belief that in the digital environment, there is no life outside media, and at the same time the human and the natural cannot be separated. Next, I will argue that *The Truman Show* can help to reframe the perception of our place in a world surrounded by media. Although the film has been traditionally linked to debates about free will, the relationship between God and man, and the influence of popular culture on contemporary societies, there is still something to add in relation to the role of the technological devices that monitor (and to some extent define) our environments. This reading highlights how the film illustrates a specific way to recognize the world around us. Finally, I will focus on the challenges that we face as a society if we consider our media life in the Anthropocene, and how films like *The Truman Show* can provide a fruitful framework in the process of making meaning of our socio-natural relationships.

Media Life and the Anthropocene

Though we may not know it, or if we do we may choose not to engage with them, media surround us and are key for the configuration of the environment in which we live. From surveillance cameras to the observation of animals, early warning technologies for earthquakes or tsunamis, facial recognition, geo-locators in our mobile devices, and technologies for medical intervention, the media environment is more and more relevant and necessary in our daily lives. Mark Deuze (2012; 2023) introduces the foundations of what it means to live a media life. Media (that is, the channels we use to communicate, what we communicate, and also the industries that enable it) are everywhere and cannot be switched off. This means that we do not live *with* media, but *in* media: the way we work, learn and define ourselves today is closely linked with this digital environment (Boczkowski and Mitchelstein 2021). Along with a media life come important communication challenges, such as keeping things private and judging the authenticity of people or messages in an information-saturated world. Even if these challenges are not new, they are more urgent in today's connected world. As Deuze puts it, "we have always used and relied on media to experience, express and comprehend our humanity, and it is up to us to take responsibility for the world we want in media" (2023, 18).

In part, these technological advances can be said to define and shape the Anthropocene, the conceptual framework that offers an explanation for the cumulative impact of human activities on planetary ecosystems. The hypothesis of the Anthropocene as a new geological epoch has been contested, but as a discursive construct and theoretical framework, it confirms

> the plausibility of a particular view of nature and the corresponding relations between the social and the natural. Even if the notion is finally not recognized by geologists or fails to capture the public imagination, the reality that it describes will not fade away.
>
> (Arias-Maldonado 2015, 76)

There is a certain ambiguity in the Anthropocene, in that "we have made the planet our own and, at the same time, it is increasingly strange to us" (Arias-Maldonado 2018, 218), that also applies to our media, since they are becoming more intimate and personal, but simultaneously unfamiliar (Deuze 2023). Both living in the Anthropocene and a life in media are invisible, irreversible, and unpredictable phenomena, even if contemporary societies are becoming aware of the situation. There is greater public interest, social concern, and scientific research focusing on both topics. However, one could argue, in the words of Ortega y Gasset (1965), that we are suffering a crisis of desire: there are many media devices around us but few common goals in front of us (Bergillos 2020). In fact, both debates have strong moral implications with social and political consequences. As a result, we need to find a sustainable reorganization of our socio-natural relations. How can media, in a broad sense, help in that process?

Studying (and Living In) Media and the Anthropocene

The notion of the Anthropocene offers a framework for the study of our socio-natural relations and encourages interdisciplinary approaches. As I have argued elsewhere, communication and media studies are well-prepared for this task, since

> media texts are fundamental in the process of raising awareness and calling for public action. On the other hand, media studies help with the challenge of dealing with the technological implications of our lives in the Anthropocene through the critical analysis of media management and the hybridization of technology, nature and human life.
>
> (Bergillos 2020, 9)[1]

For the purposes of this chapter, I emphasize that by studying (and living in) media and the Anthropocene we engage epistemic frameworks with interrelated implications and concerns. These are metaprocesses that act upon (and are acted upon by) people and their environments and in doing so they affect both the environment and people (Deuze 2012).

It is worth noting how, from this same premise, theoretical concepts are being developed that allow us to overcome the steep boundaries created by academic fields that have tended to dissociate the human and the natural, or the human and the technological. From this point of view, the study of media in the Anthropocene can benefit from approaches to naturecultures (Haraway 2003), medianatures (Parikka 2015), geomedia (Fast et al. 2019), technosphere (Haff 2014), technocene (Cera 2017), and mediacene (Gurevich 2014), etc. Most of these concepts contribute to the growing field of environmental media studies, a nascent subfield within media studies, focusing on the complex interrelations between environmental and social processes. Paralleling ecocriticism, this discipline aims to provide actionable recommendations and pays significant attention to the material aspects of digital technologies. It fosters interdisciplinary dialogues, but it adopts a more normative approach to the concept of the Anthropocene, treating it as a given rather than a subject for debate. Additionally, environmental media studies maintain a narrow understanding of media. Instead of broadening its definition, the field concentrates on the intersection of environmental issues with the production and use of new media, thereby bridging the social sciences and digital humanities. For Shriver-Rice and Vaughan (2020, 4–5), the definition of environmental media studies hinges on five concepts: a focus on media as digital screen culture; the avoidance of broad terms that take media as anything that mediates; an orientation toward action and policy recommendations; the acknowledgment of the material aspects of digital technology; and the conviction that humanity mediates the role of technology, rather than technology shaping human action.

Since screen media not only dramatize underlying socio-cultural concerns, but also serve as a way of representing the interlinked materiality and ecology of media systems, an analysis of *The Truman Show* demonstrates how the film reflects on agency, the role of media in our lives, and humans as (un)knowing protagonists of their environment. From a broader perspective, it contributes to the ongoing conceptual debate on the Anthropocene by supporting a nuanced understanding of human socio-natural relationships in an age of ubiquitous media and climate change.

The Truman Show and a Media Life

Premiering in 1998, *The Truman Show* explores the limits of reality, surveillance, and free will, among other topics. The story follows Truman Burbank, whose entire life in the town of Seahaven is broadcast as a reality television show, unbeknownst to him. The name of the protagonist ("true-man") points to the intention of the program: to live-broadcast an "authentic" life. Further, Burbank, California, is home to a vast array of media production studios. As Truman gradually discovers his fabricated existence, the film critiques media manipulation and questions the nature of such authenticity and agency. Through a satirical narrative, *The Truman Show* examines the ethical implications of voyeurism and the commodification of human life.

Beyond the enduring concern about the influence of mass media, today, 25 years after the film's premiere, *The Truman Show* can also be linked to the critique of the social and cultural dynamics that, after the adoption of mobile devices and social media platforms, condition our lives. This issue appears in the film itself, which not only represents media as a system of mass communication enterprises and powerful television formats, but also as a hybrid ecosystem of narratives, cultural conventions, automated devices, and hybridized technologies in a media-saturated environment. Like Truman's, ours is a world characterized by pervasive and ubiquitous media.

From the beginning, the film presents itself as a metafictional text with intertextual references to audiovisual content, conventions of reality television, and cinema. The initial credits introduce the protagonist of the film; not as Jim Carrey, but as Travis Burbank, the fictional character. Also, throughout the film, montage alternates a traditional cinematic narrative with images from the reality show that the film is named after. In doing so, the viewer of the film is also part of the audience of the show and similarly experiences Truman's critical awakening.

Thus, two levels at which communication and media influence contemporary societies are represented in the film. On the one hand, the film depicts the way media industries and the texts they put into circulation influence audience perceptions, and the way they make sense of how the world works. In *The Truman Show*, newspapers, misleading advertising posters, television news, and movies convince the protagonist that he lives in the right place and sustain his fears of traveling. Christoph, the creator of the television format and director of the show, refers to Plato's allegory of the cave to justify that "we accept the reality with which we are presented." In Truman's daily life, radio shows or a morning visit to the newsstand organize his routine. The media he consumes provide catchphrases that he often imitates and that he integrates in his way of speaking and behaving. Similarly, Truman's fears and concerns are born and sustained by his exposure to worrying news and alarming media content. From this perspective, the film offers a look at media that, like the field of media effects, is concerned with the way media content and consumption shape the way we see the world. Interestingly, the unsettling messages in Truman's life could be said to represent the Earth and nature as dangerous spaces. For instance, at the travel agency where Truman looks for a ticket to leave Seahaven, posters warn of problems that can occur abroad: diseases, wild animals, or lightning strikes through airplanes in mid-flight. Truman must be convinced that nature is something alien, dangerous, and uncontrollable and that it is not worth exploring.

On the other hand, media are represented as technological devices that enable different forms of communication. Truman lives unaware of the fact that he is constantly being watched and recorded. It is a media-saturated environment that resonates with today's world, full of security cameras, the mass adoption of smartphones with HD cameras, and other high-end technologies. Part of the process of invisibilization of the media has to do with their gradual mimesis with the landscape, whether through the development of wireless connection technologies or the camouflage of communication infrastructure in the natural environment. *The Truman Show* devotes a scene to explaining, through point-of-view shots and voice-over narration, how hidden cameras and microphones (in mirrors, cars, and even rings) go unnoticed in the life of the protagonist.

This infrastructural invisibility shapes Truman's media life, but also today's societies. In Parks's words, we are accustomed to knowing little about the infrastructures that surround us, even though many of us use media every day:

> A theory of media infrastructure would account not only for the bodies of actors that appear on screen but for those involved in supporting acts .... These processes are often invisible and their relation to digital humanities and media studies research is under analyzed.
>
> (2015, 370)

Likewise, as Deuze states, in a media life, "it seems our media are gradually disappearing from view while at the same time influencing our lives more and more in terms of our (real or perceived) control over them and their control over us" (2012, 60). However, if media are not recognized, they are still there and have an (invisible) impact. As Starosielski posits, the invisible is not non-existent: "In order to understand the construction of our contemporary information sphere and the infrastructure that supports it, we must move beyond conceptualizing it as naturally invisible, or only visible when it is built or disrupted" (2012, 54). In fact, media become most visible when they fail or when they stop working. In the film, it is precisely the failure of media that accompanies Truman's awareness of its existence. For instance, there is interference as he listens to the radio, a dish light from the studio drops in the middle of the street, the water hoses that cause the rain are not synchronized and fall on Truman at the wrong time, and only then does the protagonist realize the repetitive staging that accompanies his routine. As he witnesses these glitches, Truman becomes increasingly aware of the mediated environment in which he lives. Through these small 'system disruptions' and accidents he slowly understands that he is imprisoned and must do something, or find an exit. This recognition of the hidden infrastructure behind media technologies can be thought of as a way for sparking public knowledge about the materiality of wireless and other network systems. Initiatives like the critical studies of the cloud (Hogan 2023) or the Internet Tour, a project by Mario Santamaría (2024) (defined as "a journey through the physical Internet infrastructure, a tourist route of non-touristic places"), point to the need to make media visible again and reopen the questions about what we are connected to and why (Zylinska 2016).

There is a specific scene in the film where, through montage, the impact and scope of *The Truman Show* is displayed. Along with the many media professionals involved in the show, the scene introduces the resources needed to sustain it: satellites, cables, and the construction of the studio, supposedly the biggest man-made structure on Earth. While this issue is not explicitly thematized in the film, the scene suggests how these infrastructures have a huge ecological impact, thus indicating a further ambivalence of media in anthropocentric times. On the one hand, media are imperceptible, as demonstrated by the invisibility of the hidden cameras on the stage but, on the other, there is the enormous material impact embodied by the structure that houses the set, which is visible from space. Following Steffen et al., it could be argued that we are "the first generation that knows our activities influence the Earth System, and thus the first generation with the power and responsibility to change our relationship with the planet" (2011, 749). In the Anthropocene, we, like Truman, need to act in the face of an unpredictable and changing environment.

A Life in Media and the Anthropocene: Insights from *The Truman Show*

The power of media, as mentioned above, lies in their invisibility (or in their ability to go unnoticed) and in the effects they have on our lives, without us being aware of it. Indeed, their influence is reciprocal, if we understand that we also, consciously or unconsciously, redefine media through the use we make of them. In the film, Truman's actions condition the broadcast, and he develops creative ways of confronting or using the media. In the emotional journey that Truman undertakes, the film increasingly acknowledges his agency through the way he uses media, for example, when he cuts out magazines to make a portrait of Sylvia, the woman he falls in love with.

Like in other films, such as *Gattaca* (Andrew Niccol, United States, 1997) or *Free Guy* (Shawn Levi, United States, 2021), the protagonist finds the limit of his environment at the beach, in front of the sea. It is the frontier of his known world and the place where he goes to try to find answers concerning what lies beyond. In this scene, while Truman is sitting on the beach, his catharsis begins conjointly with a changing and manipulated environment. Suddenly, it starts to rain on him, and only on him, because of a technical error on the set. Despite the strangeness of the event, Truman is not afraid but rather surprised. He plays around the water jet and laughs and sings before the downpour falls on the beach. While still not aware of his role in a television show, Truman starts to embrace (environmental) change and confront his fears.

As Truman becomes more determined to leave Seahaven, Christoph, the director, decides to create extreme conditions that will force the protagonist to give up his adventure. The producers of the reality show are presented as powerful controllers of the climatic conditions affecting Seahaven. Interestingly, actor Paul Giamatti was cast as the engineer/scientist who is responsible for observing and/or manipulating the living conditions of Truman through technology and media. This actor was later typecast in this role, appearing in films such as *Paycheck* (John Woo, United States, 2003) or

Figure 6.1 A rain system malfunction at the beach. *The Truman Show* (1998).

San Andreas (Brad Payton, United States, 2015), wherein he again seems to be capable of monitoring and controlling the environment through screen technologies. Similarly, environmental historian John R. McNeill (2000) once said that we have turned the Earth into a gigantic laboratory, without being able to anticipate the outcome of an experiment still in progress. We know, like Christof, that we pushed the boundaries, and exceeded the system's limits, but what comes next?

This visual construction of the environment as something to be controlled through "geopower" introduces the problematic idea that, through science

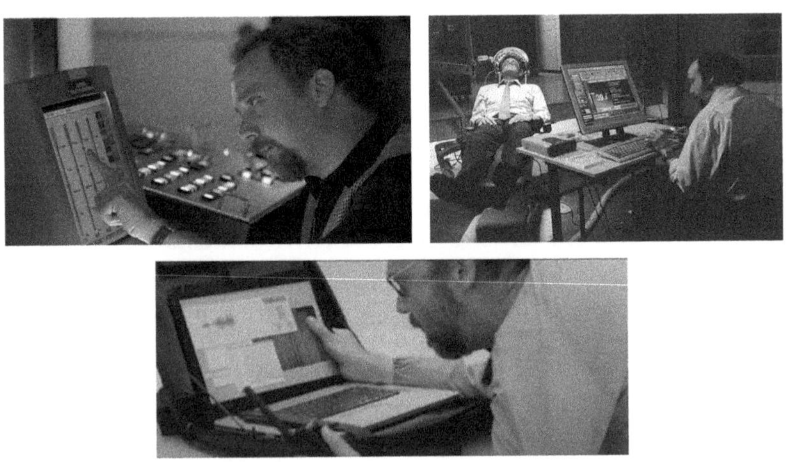

Figures 6.2, 6.3, and 6.4 Paul Giamatti as a scientist who controls men or their environment in *The Truman Show* (1998), *Paycheck* (2003), and *San Andreas* (2015).

and technology, humanity will be able to guarantee the conservation of the "Earth System" (Bonneuil and Fressoz 2016).

Christoph and the producers think they can handle the situation, but are entering an area of no return, pushing the system to its limits. The manipulation of weather and climate by man and through technology is depicted literally, and it works to symbolize our own management of the climate emergency. The image that represents the situation resembles "Show your stripes,"[2] a project by Ed Hawkings (2018), which color-codes global temperature change in the last decades in an increasingly intense red.

This functions as an apt metaphor for the ambivalence in the Anthropocene, where man is both the subject who is affected by the transformations of the planetary ecosystems and at the same time a potential creator of changes in these environments. From a critical perspective, it has been pointed out how illustrative it is that this issue has been usually introduced in cli-fi movies from the representation of the white-man-hero who saves his family and humanity; or who simply explains the problem and offers a technological solution (Leyda and Loock, 2022). It seems no coincidence that in *The Truman Show* the subject and actor of climate change (and media life) is, again, a white man.

In the film's final scene, the battle between Christof and Truman intensifies. On the one hand, Truman's catharsis drives him to the limits of survival. On the other hand, the manipulative producers force the weather system to prevent the protagonist's escape. Suddenly, Truman's boat crashes against the horizon of the television set. In a staging that pays homage to *Architecture au clair de lune* (René Magritte, 1956), natural light illuminates the human construction; a staircase leading to the exit of the studio. Finally, Truman finds

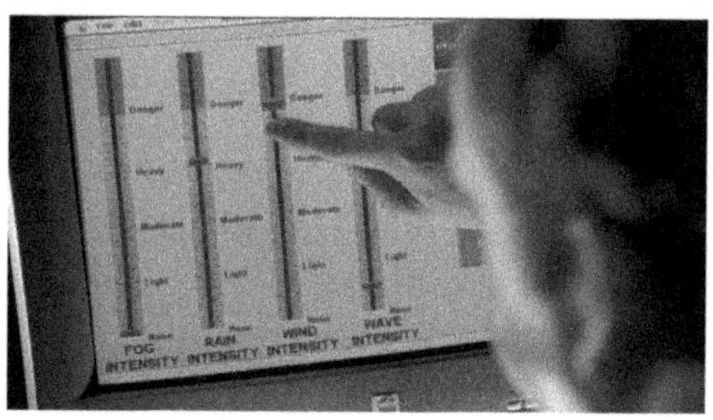

Figures 6.5 and 6.6 *The Truman Show* (1998) and "Show your Stripes" (Ed Hawkings, 2018).

out the truth and gains awareness of his imprisonment. In that specific moment, Truman, like we all must when acknowledging the Anthropocene, faces the paradox of knowing that we are powerless protagonists of our mediated lives in a climate emergency.

Authors like Jenkins (2004) have pointed out that the ending of the film would have been more powerful if it had raised the possibility that Truman, instead of leaving the set, decided to stay on his own reality show. Knowing himself to be the protagonist of the show, he could have controlled the narrative and used the media for his own purpose and following his own criteria. Accordingly, in a media life, Deuze (2012) notes, there is no exit from the studio. I argue that even if Truman left the show, he would still have to deal with the logic of media and communication technologies in the 'real world.' Also, from the environmental point of view, we can compare Truman's situation to the fact that humanity cannot escape and live outside the Anthropocene.

Conclusions

The Anthropocene urges a public debate on what kind of society we want for the future and how to achieve environmental sustainability. But it also asks for a balanced relationship with the media that shape our lives and the environment. We live in what Coeckelbergh (2022) calls a technological constellation: a life in which the media not only explain but also construct meanings about our environment. Thus, it is urgent to reconcile the ambiguity caused by our intensive use of technologies and the responsibility we must assume in the knowledge that we have a significant impact on the planetary ecosystems.

The Truman Show illustrates the paradox of being aware that the same media that shape our environment (Couldry and Hepp 2016) can also be constructed or manipulated (Enli 2015). As I have discussed, the film represents our experience of living in media (Deuze 2012), which influences and shapes our everyday life. Further, there are ways of engaging with these media so as to effect changes, even if, unlike Truman, we cannot simply escape from the set. Although the environment may be similar for everyone and the media tend to unify our experience, there are as many ways to navigate the media ocean as there are people in the world. Our relationship to media may range from total dependence to an active resistance (Syvertsen 2017) in search of disruption of media logics and can be perceived as unique and personal.

The importance of recognizing media in our lives and in our environment will contribute to more informed and meaningful discussions on how to navigate current technological and environmental challenges. Like Truman, we can only gain agency and sovereignty over our mediatized selves if we are aware of the media that surround us (and which inform and monitor the planetary ecosystems we live in), or by "gaining and raising awareness about all the subtle (and not so subtle) ways in which technologies in general and media in particular nudge and direct us, or find ways to completely rewrite the script or code to create a different (yet still mediated) world" (Deuze 2023, 223). Thus, as post-sovereign subjects[3] (Arias-Maldonado 2016), we actually gain sovereignty by realizing that our decisions and preferences are mediated processes linked to the process of mediatization.

Deuze (2023) points out that there are three options for a life in media, once we recognize we are living in them: to fight media, to surrender to a mediated existence, or to become media. While Truman in the film tries to escape, he also demonstrates that creativity is a path to agency. Arguably, there is

> a perspective beyond escape or surrender—a way of living in media with full knowledge of how media work, consistently claiming our own narrative, and choosing to enjoy life while remaining staunchly reflective and critical about their uncanny ability to mediate—to come between whatever is going on and whomever we think we are.
> (Deuze 2023, 251)

Unlike Truman, we have no way out of a media life or the (un)intentionally man-made climatic turbulences of the Anthropocene.

That is why the material implications of media in the Anthropocene need to be integrated into the debate of a media life, given that our socio-natural relations are and will be mediated by technology and its material articulations. In the film, Christoph and the producers represent a failed idea of the possible governance of the planet or the environment. The film offers the opportunity to think about our media practices from a green and environmental media perspective and can serve as a reference to establish common ground in debates on how to balance the integration of media in our lives and the responsibility of knowing that we are influential agents on planetary ecosystems.

The best way to find that common ground might be to turn to (eco)media studies and subfields like environmental media studies, that "construct alternative understandings of—and practical recommendations for—more effective communication practices, more equal resource access and potential spaces for digital resistance and community-building" (Shriver Rice and Vaughan 2020, 10). We need further environmental analysis of the logics of production, distribution, and reception of texts that have a high symbolic value and contribute to the development of a critical citizenship through the representation of our ambivalent role in the digital environment and the Anthropocene.

Notes

1 For a detailed review of the possible approaches to the study of the Anthropocene from communication science and media studies, see Bergillos (2020).
2 For further information about the project, see https://showyourstripes.info/.

3 The concept of post-sovereign subject, conceptualized by Arias Maldonado (2016), defines the paradoxical situation in which citizens feel empowered by acknowledging that as political subjects they are not as autonomous as they thought.

Works Cited

Arias-Maldonado, Manuel. 2015. *Environment and Society. Socio-natural Relations in the Anthropocene*. London: Springer.

Arias-Maldonado, Manuel. 2016. *La democracia sentimental. Política y emociones en el siglo XXI*. Barcelona: Página Indómita.

Arias-Maldonado, Manuel. 2018. *Antropoceno: La política en la era humana*. Madrid: Taurus.

Bergillos, Ignacio. 2020. "Media life in the Anthropocene." *Journal of Environmental Media* 1 (1): 27–31. https://doi.org/10.1386/jem_00004_1.

Boczkowski, Pablo J., and Eugenia Mitchelstein. 2021. *The Digital Environment. How We Live, Learn, Work, and Play Now*. Cambridge: The MIT Press.

Bonneuil, Christophe, and Jean-Baptiste Fressoz. 2016. *The Shock of the Anthropocene: The Earth, History and Us*. New York: Verso.

Cera, Agostino. 2017. "The technocene or technology as (neo)environment." *Techné: Research in Philosophy and Technology* 21: 243–81.

Coeckelbergh, Mark. 2022. "Earth, Technology, Language: A Contribution to Holistic and Transcendental Revisions After the Artifactual Turn." *Found Sci* 27: 259–270. https://doi.org/10.1007/s10699-020-09730-9.

Couldry, Nick, and Andreas Hepp. 2016. *The Mediated Construction of Reality*. Cambridge: Polity.

Deuze, Mark. 2012. *Media Life*. Cambridge: Polity Press.

Deuze, Mark. 2023. *Life in Media. A Global Introduction to Media Studies*. Cambridge: The MIT Press.

Enli, Gunn. 2015. *Mediated Authenticity. How the Media Constructs Reality*. New York: Peter Lang.

Fast, Karin, Emilia Ljungberg, and Lotta Braunerhielm. 2019. "On the social construction of geomedia technologies." *Communication and the Public* 4: 89–99.

Festival de Cannes. 2022. "The Official Poster of the 75th Festival de Cannes." https://www.festival-cannes.com/en/press/press-releases/the-official-poster-of-the-75th-festival-de-cannes/.

Gurevich, Leon. 2014. "Google Warming: Google Earth as eco-machinima." *Convergence: The International Journal of Research into New Media Technologies* 20: 85–107.

Haff, Peter. 2014. "Humans and technology in the Anthropocene: Six rules." *The Anthropocene Review* 1: 126–36.

Haraway, Donna J. 2003. *The Companion Species Manifesto: Dogs, People, and Significant Otherness*. Chicago: Prickly Paradigm Press.

Hogan, Mel. 2023. "Environmental media in the cloud: The making of critical data center art." *New Media & Society* 25 (2): 384–404. https://doi.org/10.1177/14614448221149942.

Jenkins, Henry. 2004. "The Cultural Logic of Media Convergence." *International Journal of Cultural Studies* 7 (1): 33–43.

Leyda, Julia, and Kathleen Loock. 2022. "Climate Fictions, Dystopias and Human Futures (A video essay by Julia Leyda and Kathleen Loock)." https://www.youtube.com/watch?v=YRR-Ohcz4y8&t=589s.

McNeill, John R. 2000. "Our Gigantic Experiment With Planet Earth." *The Chronicle of Higher Education*, May 19, 2000. https://www.chronicle.com/article/our-gigantic-experiment-with-planet-earth/.

Ortega y Gasset, José. 1965. *Meditación de la técnica*. Madrid: Espasa Calpe.
Parikka, Jussi. 2015. *A Geology of Media*. Minneapolis: University of Minnesota Press.
Parks, Lisa. 2015. "Stuff you can kick: Toward a theory of media infrastructures." In *Between Humanities and the Digital*, edited by Patrik Svensson and David T. Goldberg, 355–73. Cambridge: MIT Press.
Rust, Stephen, Salma Monani, and Sean Cubitt. 2013. *Ecocinema Theory and Practice*. New York: Routledge.
Santamaría, Mario. "Internet Tour." Accessed May 21, 2024. https://internetour.com/.
Shriver-Rice, Meryl, and Hunter Vaughan. 2020. "What is environmental media studies?" *Journal of Environmental Media* 1: 3–13. https://doi.org/10.1386/jem_00001_2.
Starosielski, Nicole. 2012. "'Warning: do not dig': negotiating the visibility of critical infrastructures." *Journal of Visual Culture* 11 (1): 38–57. https://doi.org/10.1177/1470412911430465.
Steffen, Will et al. 2011. "The Anthropocene: from global change to planetary stewardship." *Ambio* 40 (7): 739–761. https://doi.org/10.1007/s13280-011-0185-x.
Syvertsen, Trine. 2017. *Media Resistance. Protest, Dislike, Abstention*. London: Palgrave Macmillan.
Zylinska, Joanna. 2016. "Cables, cameras and other fossils: Media pasts and media futures." In *6th ECREA European Communication Conference: Mediated (Dis)Continuities: Contesting Pasts, Presents and Futures*, Prague, November 9–12, 2016.

Part 2

More-Than-Human Kinship, Hybridity, and Monstrous Alliances

7 Collaborative Making, Not Taking

Nova Paul Exposes Cinema's Material Roots

Missy Molloy

Introduction

In a conversation about her film installation *Ngā Pūrākau Nō Ngā Rākau* (Aotearoa New Zealand, 2023), Nova Paul described the frustrating process of trying to film three rākau ("tree" in te reo Māori) that hold great significance for her hapū (te reo Māori for a subtribe of an iwi, the larger groups that comprise Māori society). Paul recounts "wast[ing] a day in the lab," surprised not to "get any images" despite shooting in "perfect lighting conditions" (Paul 2023). Then, after discussing the situation with her mentor Dinah, Paul was guided to film "the matua rākau in the ngāhere" (elder trees in the forest) instead. In that case, and despite "terrible" lighting conditions that required Paul to set the aperture of her 16mm camera "wide open" and shoot near-blind (due to limited light), "the footage turned out the best," manifesting an "evocative" quality that "opens up," in Paul's (2023) words, "te ao mārama" (literally "the world of light" in te reo),[1] which is the overarching goal of her rigorously material, committedly retro film practice [Figure 7.1]. In "Ngā Pūrākau Nō Ngā Rākau: Stories from Trees," an article Paul wrote with Tessa Laird to provide insight into the making of her short films, *Rākau* (Aotearoa New Zealand, 2022) and *Hawaiki* (Aotearoa New Zealand, 2022), both of which feature as part of the *Ngā Pūrākau Nō Ngā Rākau* installation, Paul connects the emphasis on light in the Māori creation story—in which Tāne, "God of the Forest," pushes apart his "sky father" Ranginui and "earth mother" Papatūānuku to reveal "the world of light" (Paul and Laird 2023, 7)—to film, "a medium of light," with Paul and Laird remarking, "Nova's films have always been woven, meshing together layers of colour, time, place, and perspective. But in attending to light, we cannot ignore the foundational role of darkness, either in cinema, or in the growth of plants and people" (8).

Paul's multimedia practice is aptly described as "encompass[ing] early cinematic film processes, experimental film practices and weaving traditional stories and mātauranga Māori [which "refers to Māori knowledge in its widest and broadest terms"[2]] to explore the poetics and politics of everyday self-determination and place, ultimately envisioning tino rangatiratanga—Māori

DOI: 10.4324/9781003468028-9

Figure 7.1 The Pūriri tree is central to *Hawaiki* (2022) and to Nova Paul's film installation, *Ngā Pūrākau Nō Ngā Rākau* (2023), which I experienced at Te Whare Toi/City Gallery Wellington on October 9, 2023, when I took this photograph.

sovereignty."[3] This chapter will present Paul's cinema, in particular *Ngā Pūrākau Nō Ngā Rākau* ("Stories from Trees"), as an illustration of what Katarzyna Paszkiewicz refers to as "affective and new materialist" (2021, 5) cinema, which, she suggests, can allow us to "resituate" ourselves in what is known today as the Anthropocene (4). I will also show that *Ngā Pūrākau Nō Ngā Rākau* cultivates "identification with the nonhuman" in a direct rather than suggestive or symbolic manner, thereby evoking what Paszkiewicz describes as "a bodily, rather than purely visual, reciprocity with the world [that] can be particularly fruitful for thinking about the nonhuman" (11), and that it does so in a manner illuminated by the Māori "master concept" of whakapapa (Stewart 2021, 85). My analysis of Paul's work will emphasize that the material density of *Ngā Pūrākau Nō Ngā Rākau*'s footage and audio in combination with the design of its installation evokes embodied encounters between human and nonhuman subjects on- and off-screen in the exhibition space.

In order to claim that Paul's work is in line with an "affective and new materialist" strand of the cinema that critically responds to the discourse on the Anthropocene, I must contend, first, with the skepticism many Indigenous scholars have expressed regarding both the Anthropocene—"an unofficial unit of geologic time used to describe the most recent period in Earth's history when human activity started to have a significant impact on the planet's climate and ecosystems"[4]—and new materialisms, "a theoretical turn away from the persistent dualisms in modern and humanist traditions" that "reposition[s] ... the human among nonhuman" agents and "advocate[s for] a critical materialist attention to the global, distributed influences of late capitalism and climate change" (Sanzo 2018). As outlined below, Indigenous criticism of these closely allied fields of scholarly inquiry is consistent and persuasive. That said, both Indigenous and Anthropocene studies evince a strong commitment to addressing the urgent environmental crises that Kali Simmons itemizes as: "Destruction of homelands. Loss of kinship species. Exposure to deadly contaminants. Mass extinction. Transformed lifeways" (2019, 174). For Paul, the story of the three rākau communicates the fact that *Ngā Pūrākau Nō Ngā Rākau* is "about relationships with and within the environment," and, as such, "the tree would give [her] an image, [she] wouldn't take the image" (Paul 2023). This story and the film installation it informed explicitly employ a Māori lens to oppose "the colonial gaze and extractive processes that often happen ... in moving-image making" by using the cinematic "medium of light" to receive rather than take images (Paul 2023). Moreover, *Ngā Pūrākau Nō Ngā Rākau* and its paratexts model ways of rethinking and (in Paszkiewicz's words) "imagining our contact with the nonhuman ... to dissolve the categorical divide between the human and the nonhuman" (Paszkiewicz 2021, 3). Hence, I argue that Indigenous-oriented film practice can shed much needed light on the dynamic between human and nonhuman environments and compellingly enacts the profound perspectival shifts that many scholars have sought under the rubrics of the Anthropocene and new materialisms. My account of Paul's work will thus stress the value of centralizing Indigenous concepts and aesthetics in environmental theory and action in order to strongly urge scholars engaging with the Anthropocene and new materialisms to recalibrate their critical endeavors to explicitly acknowledge Indigenous contributions to areas of overlapping concern.

Indigenous-Oriented Objections to the Anthropocene and New Materialisms

Indigenous critique of ideas central to theories of the Anthropocene and new materialisms abounds. The criticism condenses on three points that I will emphasize before bringing them to bear on my reading of Paul's work: Indigenous scholars have (1) disputed the newness of these theories and their takes on environmental crisis and the human/nonhuman dynamic; (2) subjected the timelines on which theories of the Anthropocene are (and have been) based to close scrutiny; and (3) warned the proponents of such theories that they risk erasing the histories, experiences, and perspectives of Indigenous peoples. All three points are convincingly presented by, among others, Brendan Hokowhitu, Kyle Powys Whyte, and Kali Simmons, whose scholarship on the kindred theories I draw upon below (Hokowhitu 2022; Whyte 2017; Simmons 2019).

A number of Indigenous scholars have successfully raised the question of whether these related theoretical positions are in fact new. In "The Emperor's 'New' Materialisms: Indigenous Materialisms and Disciplinary Colonialism," Hokowhitu delivers sharp commentary on "New Materialisms" and related theories (including those associated with posthumanism and the Anthropocene) on the ground that

> the theoretical drivers and philosophical currents that hastened the theoretical turn away from the multiple binaries and humanist traditions underpinning Western thought since the Enlightenment, at least, were already well and truly ensconced in our discipline and the fabric of our cultures and epistemologies.
>
> (2022, 131)

In rejecting new materialisms' novelty claim, Hokowhitu references Métis scholar Zoe Todd's dismay upon witnessing a 2013 lecture by Bruno Latour on "the climate as a matter of 'common cosmopolitical concern'" wherein she waited in vain for the speaker to "credit Indigenous thinkers for their millennia of engagement with sentient environments, with cosmologies that enmesh people into complex relationships between themselves and *all* relations, and with climates and atmospheres as important points of organization and action," while the celebrated theorist neglected to mention a single Indigenous thinker (Todd 2016, 7–8).[5] Hokowhitu ultimately concludes that "the nomenclature of 'new' is simply offensive in the broader realm of multiple realities because it's [sic] claims to temporal ownership of ideas that already existed in multiple Indigenous philosophies" evince a delusional colonial mentality that is incapable of recognizing ideas "not delivered into this world by European/US reason" (2022, 132).

This rejection of novelty claims by allied theoretical "turns" self-branded as new at the expense of time-honored Indigenous knowledge directly relates to my second major thread of critique here, which identifies major flaws in timelines of the Anthropocene that fail to register environmental catastrophes that have already happened to and been theorized by Indigenous peoples. In "Reorientations; or, An Indigenous Feminist Reflection on the Anthropocene," Kali Simmons suggests that a large part of Indigenous scholars' skepticism about the Anthropocene lies in their belief that the environmental problems this academic "framework" aims to address have long been the foci of Indigenous communities—and out of necessity rather than choice. Per Simmons, "Indigenous peoples have already faced catastrophic violence, the loss of relationships, and the fundamental alteration of their ways of life to survive in spaces that are physically, emotionally, and spiritually toxic" (2019, 174–175). In "Indigenous Science (Fiction) for the Anthropocene: Ancestral Dystopias and Fantasies of Climate Change Crises," Kyle Powys

Whyte raises similar objections to how some accounts of the Anthropocene approach environmental crisis as "impending," whereas "the hardships many nonIndigenous people dread most of the climate crisis are ones that Indigenous peoples have endured already due to different forms of colonialism: ecosystem collapse, species loss, economic crash, drastic relocation, and cultural disintegration" (2017, 225–226).

In advancing this position—that many Indigenous people have already experienced environmental crisis at the level of severity and intensity that theories of the Anthropocene tend to frame as either relatively recent or as "impending"—Whyte draws on Heather Davis and Zoe Todd's influential article, "On the Importance of a Date, or Decolonizing the Anthropocene," which effectively argues that "the question of *when*" "the current geological time frame should be considered the Anthropocene" has "political consequences" that should be considered alongside the debate about "*whether or not*" it should be accepted as a legitimate geological epoch (Davis and Todd 2017, 762). On a related note, Hokowhitu reiterates Simmons's and Whyte's claim that "Indigenous thinkers have been doing this work for millennia" (Hokowhitu 2022, 144), while delving into the political implications raised by Davis and Todd. Thus, he concludes that not only have the timelines for framing the catastrophic dynamic between certain humans' behaviors and their consequences been revealingly flawed, but also that this temporal error signifies an oversight that is at best careless, but is, regardless of intention, colonialism manifesting in yet another guise (thus his chapter title, "The Emperor's 'New' Materialisms"):

> This chapter has enabled me to engage with some of the New Materialisms /Posthumanism literature .... In its forgetfulness it repossesses the 'rights' to knowledge .... The fact that Indigenous Studies and peoples have largely been absent or invisible in these burgeoning discussions is symptomatic of Western theory's *pathology* and *dis*ease.
>
> (143–144)

Addressing this debatably "patholog[ical]" oversight, Indigenous scholars have repeatedly posed the question of whether these purportedly novel categories of theory are even necessary given the existence of robust scholarly inquiries already rooted in postcolonial and Indigenous critiques of Western binaristic thinking, in particular the conceptual split of "human" from "nature" that enabled capitalistic exploitation of humans and nonhumans alike, which set in motion what is now widely discussed as a climate crisis of global proportions.

What, then, are the effects of the major temporal flaws and revealing omissions that Indigenous scholars have consistently drawn attention to, and can said flaws be adequately dealt with to salvage these emerging and fast-solidifying theoretical fields that are attracting scholars in ever-increasing numbers? In my view, these theories (and those using them) need to seriously reckon with the possibility that their founding premises might have the effect, intended or not, of "eras[ing] Indigenous peoples' perspectives on the connections between climate change and colonial violence," as per Whyte (2017, 225). Simmons delivers the same warning, likewise in terms of erasure but with an emphasis on Indigenous feminisms:

> To approach the Anthropocene from a position that does not remember the history of empire is a continuation of the system of settler colonialism that erases indigenous peoples so the settler nation can be imagined as empty and occupiable. Indigenous feminisms warn against responses to the Anthropocene that reinscribe this violent erasure, taking as their starting point the intersections of empire, industrial capitalism, and heteropatriarchy.
>
> (2019, 175)

Davis and Todd also lament the erasure that would result from mistakenly dating, historicizing, and theorizing irrevocable environmental harm when they write, "without recognizing that from the beginning, the Anthropocene is a universalizing project, it serves to re-invisibilize the power of Eurocentric narratives, again re-placing them as the neutral and global perspective" (2017, 763). This review of Indigenous-centered scholarly critique of the Anthropocene and new materialisms demonstrates remarkable consistency regarding Indigenous scholars' objections to these kindred branches of contemporary, environmentally engaged theory. The fact that, in many cases, the same words are used to express closely aligned criticism indicates that these interdisciplinary scholars are well aware of each other's work. The question of why non-Indigenous scholars seem less cognizant of (or engaged with) this work is one that advocates of these theories should take up in order to redress the valid points summarized above.

Exhibiting Cinema's Material Dynamism: *Ngā Pūrākau Nō Ngā Rākau*

Despite the credibility of the objections Indigenous scholarship has raised against the notion of the Anthropocene, Simmons states that the debate on the Anthropocene nonetheless has the potential to "bring attention to the violence indigenous people have suffered and continue to resist" but only if "three key concepts," "time, contamination, and kinship," are "radical[ly] reorient[ed]" in approaching the hypothetical historical epoch (2019, 175). Of these three, the first and third illuminate Nova Paul's unique filmmaking process, which revives early cinema techniques to reflect te ao Māori (a Māori worldview), as elaborated on in Paul and Laird's article ("Ngā Pūrākau Nō Ngā Rākau: Stories from Trees"). *Ngā Pūrākau Nō Ngā Rākau* is a "moving image installation [that] weaves together imagery and sound, drawing relational connections through a visual vocabulary anchored in rākau (trees) and the ngāhere (forest)."[6] The installation, comprised of four simultaneous large-scale

Figure 7.2 *Ngā Pūrākau Nō Ngā Rākau*'s four-screen projection envelopes the visitor in an array of natural environments unified by a focus on rākau but including many other elements, such as mountains and the sea.

projections [Figure 7.2], animates an alternative temporality that reflects "First Peoples' ontologies in which time is not linear, but a meshwork of past, present, and future" (Paul and Laird 2023, 4).[7]

In *Ngā Pūrākau Nō Ngā Rākau*, a Māori approach to time interacts with Paul's rigorously material process to produce an experience of whakapapa, or "grounded, ancestral, and material relations" (Paul and Laird 2023, 6). Te Whare Toi/City Gallery Wellington's curatorial statement describes Paul's process as follows: "Turning to rākau [trees] as a physical component in her filmmaking process, Paul created a plant-based developer incorporating leaves and bark of the trees that she films, using it to hand process black and white sixteen-millimetre film."[8] Elaborating on this process in their article on the installation, Paul and Laird assert that

> Nova's films … exceed the indexical; the fact that the images are made with 'tree juice' and therefore carry DNA material, imbues them with whakapapa [a foundational Māori concept that I expand on below]. They are not remnant traces, but living relations.
>
> (2023, 5)

This assertion recalls Simmons' broader claim about Indigenous approaches to kinship, which she considers pivotal to reorienting the framework of the Anthropocene:

> For indigenous peoples, kinship is not simply shorthand for family or peers; it is an entire system that structures how one moves through the world. This means that other forms of life subsist outside the narrow purview of the human and that these kin and kinship systems cannot be forsaken.
>
> (2019, 178)

In my opinion, Paul's revival of early cinematic practices is not anachronistic, but rather should be understood as repurposing analog filmmaking to express a Māori view of being and relating in the world, which involves a past-oriented approach to time that is not linear in a Western sense (as Georgina Tuari Stewart explains, "we walk backwards into the future" with the past in view [2021, 66]). As Paul describes it, "Plant-based developers are an early 19th century experimental practice and I've basically refined the processes and developed a methodology to get these beautiful quality images, produced by tree juice" (Higham 2023). Paul's revitalization of early cinema processes reflects a Māori sense of time (as past-facing) and relating, wherein she captures and develops images in concert with the trees in the forest—her elder kin.

Ngā Pūrākau Nō Ngā Rākau's stance in relation to trees and other forms of nonhuman life recalls Manuel de Landa's concluding statement in a 1995 article widely considered foundational to new materialisms:

> Living creatures, according to this stance, are in no way 'better' than rocks. Indeed, in a nonlinear world in which the same basic processes of self-organization take place in the mineral, organic and cultural spheres, perhaps rocks hold some of the keys to understand sedimentary humanity, igneous humanity and all their mixtures.
>
> (1995)

Figures 7.3 and 7.4 *Rākau* (2022) can be interpreted, per Williams, as a tree "imag[ing] itself" or in line with new materialist notions of "matter ... simply *speaking*." For Nova Paul, close attention to rākau illuminates storytelling from a Māori perspective, as well as Māori "knowledge systems."

Yet this founding document demarcating a theoretical turn is entirely absent of any engagement with Indigenous ideas or scholarship, thus confirming Indigenous scholars' criticism of new materialisms as suspiciously selective in acknowledging conceptual precursors of its fundamental premises. In fact, Paul's approach to documenting the material world—trees, in particular—offers a model for circumventing anthropocentric ways of seeing in favor of close attention to the rich layers of information and knowledge that become visible in its stead [Figures 7.3 and 7.4]. In a conversation recorded in 2022, "The DNA of Film," Mark Williams states that *Rākau*

> is conceived as a completely cyclic event where a 16mm film of a rākau images itself thanks to a developer solution made from the tree's discarded leaves or the leaves you pick, which you apply to the film negative and create the image.
>
> ("The DNA of Film," 2022)

Williams's description of *Rākau*, as the tree "imag[ing] itself," lines up with Vicki Kirby's take on new materialisms, in which "matter appears as something that is not only spoken about or spoken with, but rather as itself simply *speaking*" (Van der Tuin and Dolphijn 2010, 165). When Williams asks Paul about her intentions in relation to the rākau, Paul answers:

> when we're talking about the DNA of film, from a Māori point-of-view, we're talking about whakapapa, which are the connections that we have ancestrally, but to everything, the taiao [environment] and the world .... The base of our trees is our stories ... and here, the tree[s], our relatives and elders [hold the answers to addressing climate change].
>
> ("The DNA of Film," 2022)

In framing her filmmaking practice as receiving images of trees in collaboration with them, Paul emphasizes the etymology of the te reo Māori word for storytelling, "pūrakau," which combines "base of" with "tree." According to Paul, this unique method of telling stories through film processes that materially involve the tree places "a Māori way of seeing the world and our knowledge systems at the foreground of our image-making," one that *Ngā Pūrākau Nō Ngā Rākau* expands upon in its orchestration of *Rākau* in concert with multiple audiovisual streams in a shared space.

Specifically Māori concepts of time are relevant to the temporality *Ngā Pūrākau Nō Ngā Rākau* exhibits, as are other Indigenous theories of time, which Whyte (2017) collectively terms "spiraling time." In *Māori Philosophy: Indigenous Thinking from Aotearoa*, Georgina Tuari Stewart explains a Māori view of past, present, and future as follows:

> In comparison with Western culture, Māori thinking is oriented to the past. In te reo Māori the phrase 'i muri' means 'before' in both senses: past time and in front of us ... the past is before us because we can see

> it; we walk backwards into the future since we cannot look and see what it will bring. This orientation to the world encourages us to reflect on and learn from the past.
>
> (2021, 66)

The blatant disconnect between Western and Indigenous understandings of time undoubtedly helps fuel the misgivings Indigenous scholars have expressed about the Anthropocene and related theories that aspire to intervene in a catastrophic present or near future. For example, a *Guardian* article published at the time of writing (May 8, 2024) is headlined "World's top climate scientists expect global heating to blast past 1.5C target," and, three paragraphs in, reads, "Many of the scientists envisage a 'semi-dystopian' future, with famines, conflicts and mass migration, driven by heatwaves, wildfires, floods and storms of an intensity and frequency far beyond those that have already struck."[9] Whyte would associate these allegedly scientific statements with what he refers to as "ancestral [dystopian] fantasies" passed down via settler colonialism (2017, 237–238). Such fantasies, he argues, oppose Indigenous approaches to time, which tend to evoke "a spiraling temporality ... in which it makes sense to consider ourselves as living alongside future and past relatives simultaneously as we walk through life" (228–229). In comparison, settler-colonial approaches to time promote the avoidance of problems until they manifest as crises, which are then spectacularized in a manner that perpetuates ignorance of the past (because a clear view of the past would force engagement with impending crisis before it graduates to a critical stage). In *Ngā Pūrākau Nō Ngā Rākau*, spiraling time is evident in the simultaneous projection of footage of "the matua rākau"—which Nova approached as a living relative and an ancestor,[10] in close-up, hand-processed black-and-white footage—alongside color footage[11] from the short film, *Hawaiki*, in which a group of young children talk, play, sing, and weave while draped on and around the tree [Figures 7.5 and 7.6]. Thus, while

Figures 7.5 and 7.6 These representative stills from two of *Ngā Pūrākau Nō Ngā Rākau*'s projections demonstrate the diversity of the tree imagery that unifies the installation, which characterizes trees in dynamic relationships with their human and nonhuman environments in a temporality that opposes Western linearity.

the two have distinct visual and atmospheric elements, the installation presents them alongside one another in an assertion of nonlinear temporality that invites viewers into a dynamic experience of whakapapa ("grounded, ancestral, and material relations," per Paul and Laird [2023]).

In a conversation about *Ngā Pūrākau Nō Ngā Rākau* (with Karl Chitham), Paul articulates a kaupapa Māori perspective on the link between the climate crisis and kaitiakitanga, a Māori concept that encompasses "the conservation, replenishment and sustainability of the taiao, the environment."[12] Per Paul,

> The reason that we've got a climate crisis ... is because people forget that they've actually got a deep relationship with the land. We're not kaitiaki [guardian] of it, it's kaitiaki of us. I'm really interested in how an image can create relationships with the person that's sitting in the space experiencing the film ... everything we see in the environment is an indicator of this reciprocity.
>
> (2023)

In offering *Ngā Pūrākau Nō Ngā Rākau* as an opportunity to experience whakapapa in action, Paul hopes that this embodied experience of kinship will inspire a sense of reciprocity and care for the more-than-human world, which includes living beings, human among them. She further contextualizes her work on *Ngā Pūrākau Nō Ngā Rākau*, and the work she hopes it performs in the exhibition space, as supporting broader efforts to assert Māori sovereignty in order to properly practice kaitiakitanga in accordance with whakapapa:

> Everything, everything, is given to that wai māori, that freshwater space, to assert our tino rangatiratanga mana motuhake of our awa and our puna [sovereignty and authority over freshwater and its sources] ... protecting the freshwater from things like water bottling plants and agricultural irrigation has a deep impact on daily life of the haukāinga [homeland]. In a way, the reason the films look the way they do is to whakamana [acknowledge or validate] all of that mahi [work] .... It's context for the work; and the context is whakapapa.
>
> (2023)

Tuari Stewart describes "whakapapa" as a

> master concept in Māori philosophy because it not only structures the content of knowledge about the world (beginning with the creation stories and continuing to include all the physical elements of the natural world) but also provides an ethical framework based on kinship with/in the world.
>
> (2021, 59)

Paul aims to stimulate this sense of "kinship with/in the world" through her films, which "bring together the ihi (essential force, personal magnetism), wehi (a response of awe in reaction to ihi) and wana (exhilaration) that is inherent in Māori performance" (Higham 2023). In a sense, *Ngā Pūrākau Nō Ngā Rākau* offers an opportunity to appreciate the "full range of meanings" expressed through the concept of whakapapa, which Tuari Stewart literally translates as "'layer-upon-layer' ... 'to make layers' or something like 'generative'" (2021, 85).

The relevance of whakapapa to the material relationships that *Ngā Pūrākau Nō Ngā Rākau* facilitates between cinematic technologies, the nonhuman world, and embodied humans (including Paul, sensitively wielding her camera as she moves through the ngāhere [forest], and gallery visitors like me, immersed in the sounds and images emitting from the four screens in the carefully curated installation space) seems clear. Yet the link between the installation and what has been theorized as "affective and new materialist" cinema is also evident. Consider the connection Paszkiewicz, building on phenomenologically informed film theory, draws between affective cinema and "ecologically oriented approaches to cinema" (2021, 9):

> Formulating film viewing in terms of modes of embodied spectatorship, affective connectivity or a phenomenological sharing of lived time seems apposite when addressing the nonhuman worlds because, as [Jennifer] Barker notes, the film and the viewer's body can be thought of "as intimately related but not identical, caught up in a relationship of intersubjectivity and co-constitution."
>
> (Paszkiewicz 2021, 14–15; Barker 2009, 12–13)

This understanding of film viewing aligns with the ideas about whakapapa canvassed above (including Paul's regarding her authorial intentions in undertaking the project), and both resonate with *Ngā Pūrākau Nō Ngā Rākau*'s orchestration of time and space to enable generative encounters between a variety of bodies (cinematic, nonhuman, human, technological). Moreover, these clearly heed "the ecological call 'to stretch our modes and sites of awareness, sensitivity, and attachment' to the world by artistic, and more particularly cinematic, means" (Connolly in Paszkiewicz 2021) which, according to Paszkiewicz, "has become acute in the Anthropocene" (2021, 16–17). For Indigenous artists and thinkers, this ecological call seems to have been a consistent and pressing concern, meaning they did not necessarily require any theoretical turn, yet Paul's cinema reveals the remarkable capacity of the medium to expand our perception of the material connections that enmesh us in the natural world, which we are part of rather than separate from (the installation format enhancing this perception). When I stood in the installation space surrounded by projections of Paul's films, I was able to simultaneously experience multiple environments, temporalities, and ways of being—for instance, the elder rākau footage [Figures 7.1 and 7.3], which moves slowly and complements the saturated analog photography; alongside extreme long shots of mountains and moving water [Figure 7.2] that juxtapose the close-ups that dominate the hand-processed tree imagery; and finally, the children playing in and around the tree [Figure 7.4], which injects an everyday quality into *Ngā Pūrākau Nō Ngā Rākau*'s environment. Scholars energized by concepts of the Anthropocene and new materialisms have much to learn from cinema like *Ngā Pūrākau Nō Ngā Rākau*, which enlivens principles central to these theories while modeling the perspectival shifts necessary to effectively support the shared goal of recalibrating encounters between humans and nonhumans to promote "ethical relationships between people, and between humans and other living and non-living things" (Stewart 2021, 85).

Conclusion

This chapter takes up Indigenous criticism of theories of the Anthropocene and new materialisms to ground the discussion of Nova Paul's *Ngā Pūrākau Nō Ngā Rākau* on that critique while also suggesting that her work provides an aperture for considering the relevance of these debatably new theories to Indigenous cinema, which prompts material and affective encounters that underscore the reciprocity at the heart of the natural world, humans and nonhumans included. I presented Paul's recent film installation (which, although explicitly oriented to mātauranga Māori, illustrates characteristically Indigenous approaches to time, space, materiality, and kinship) to stress the value and necessity of more explicit engagement with Indigenous theories, worldviews, and aesthetics in pursuing the goals articulated via these theoretical "turns," which aim to reconfigure the relationships between humans and their environments and to break patterns of material destruction that distinguish global capitalism in what is often referred to as a "late" stage.

The environmental problems these theoretical frameworks aim to address have long been the foci of Indigenous communities—and out of necessity rather than choice. Those genuinely grappling with these bodies of theory and the ecological realities they respond to likely do not intend to reproduce power inequities by exercising a type of intellectual freedom that is in fact a privilege bestowed by colonialism to settler descendants, and one that Indigenous people often remain excluded from (e.g. recall the Zoe Todd/Bruno Latour incident also mentioned by Hokowhitu). Regardless, Hokowhitu reminds readers that neo-colonial threats and aggressions demand constant attention on the part of Indigenous communities, which means that Indigenous scholars do not have the same range of choice in their critical pursuits that many non-Indigenous scholars enjoy:

> Given the immediacy of neo-colonialism, our scholarship has tended to remain in the juridical-political human domains in support of Indigenous communities' steadfast focus on the survivance of their 'human rights,' lands, and cultures via notions of peoples, nation and tribal sovereignty.
> (2022, 132–133)

This chapter has queried the reticence on the part of both Indigenous and non-Indigenous scholars to involve Indigenous scholarship and concepts in relevant theoretical enterprises that are rapidly gaining momentum. The above provides some useful information regarding this reticence, but in closing, I emphasize that a chasm seems to exist that discourages engagement between the two parties despite their shared commitment to environmental care—and perhaps there is a sense on the part of non-Indigenous scholars that their participation is not wanted in the space of Indigenous studies, and vice versa with regard to theories of the Anthropocene and new materialisms (especially given the documented tendency of the latter to ignore Indigenous ideas that preceded and are obviously relevant to their critical enterprises). Creative solutions for overcoming this rift seem well worth investing in so that communities genuinely committed to pursuing less-destructive worlds to inhabit now and in the future can collaborate to make them possible.

Acknowledgement

Missy Molloy would like to warmly thank Nova Paul for her generosity in sharing and discussing her work.

Notes

1 Te reo for the world, or "the world of light" (Paul and Laird 2023, 7).
2 https://e-tangata.co.nz/comment-and-analysis/understanding-matauranga-maori/.
3 https://artnow.nz/events/making-spaces-with-nova-paul.
4 https://education.nationalgeographic.org/resource/anthropocene/.
5 Quoted in part by Brendan Hokowhitu (2022, 132).
6 https://citygallery.org.nz/exhibitions/nova-paul-nga-purakau-no-nga-rakau/.
7 Paul and Laird made this statement about Nova's 2010 film, *This is Not Dying*, but it also applies to the 2023 installation, *Ngā Pūrākau Nō Ngā Rākau*.
8 https://citygallery.org.nz/exhibitions/nova-paul-nga-purakau-no-nga-rakau/.
9 https://www.theguardian.com/environment/article/2024/may/08/world-scientists-climate-failure-survey-global-temperature.
10 In Nova's words, "the trees are our whanaunga and tūpuna, they're our tuākana, and we are in relationship with everything in the taiao." See: https://artnews.co.nz/nova-paul-you-point-to-the-world/.
11 The 16 mm colour footage was lab, not hand-processed.
12 https://www.sciencelearn.org.nz/resources/2544-understanding-kaitiakitanga.

Works Cited

Barker, Jennifer M. 2009. *The Tactile Eye: Touch and the Cinematic Experience*. Berkeley: University of California Press.

Davis, Heather, and Zoe Todd. 2017. "On the Importance of a Date, or Decolonizing the Anthropocene." *An International Journal for Critical Geographies* 16 (4): 761–780.

De Landa, Manuel. 1995. "The Geology of Morals: A Neo-materialist Interpretation." *Virtual Futures* 95. http://www.t0.or.at/delanda/geology.htm.

Higham, Tim. 2023. "A Story of the Land." *Aotea: Great Barrier Island*, January 2023. https://www.greatbarrier.co.nz/a-story-of-the-land/.
Hokowhitu, Brendan. 2022. "The Emperor's 'New' Materialisms: Indigenous Materialisms and Disciplinary Colonialism." In *Routledge Handbook of Critical Indigenous Studies*, edited by Brendan Hokowhitu, Aileen Moreton-Robinson, Linda Tuhiwai-Smith, Chris Andersen, and Steve Larkin, 131–146. London: Routledge.
Paszkiewicz, Katarzyna. 2021. "Cinema and Environment: The Arts of Noticing in the Anthropocene." *Res Rhetorica* 8 (2): 2–21.
Paul, Nova. 2023. "Nova Paul: 'You Point to the World.'" *Art News Aotearoa*, November 9, 2023. https://artnews.co.nz/nova-paul-you-point-to-the-world/.
Paul, Nova, and Tessa Laird. 2023. "Ngā Pūrakau No Ngā Rākau: Stories from Trees." *Philosophies* 8 (1). https://doi.org/10.3390/philosophies8010015.
Sanzo, Kameron. 2018. "New Materialism(s)." *Critical Posthumanism: Genealogy of the Posthuman*, April 25, 2018. https://criticalposthumanism.net/new-materialisms/.
Simmons, Kali. 2019. "Reorientations; or, an Indigenous Feminist Reflection on the Anthropocene." *Journal of Cinema and Media Studies* 58(2): 174–179.
Stewart, Georgina Tuari. 2021. *Māori Philosophy: Indigenous Thinking from Aotearoa*. London: Bloomsbury Academic.
"The DNA of Film." 2022. *CIRCUIT Artist Moving Image*, December 7, 2022. https://soundcloud.com/circuit-2/episode-109–the-dna-of-film.
Todd, Zoe. 2016. "An Indigenous Feminist's Take on the Ontological Turn: 'Ontology' is just Another Word for Colonialism." *Journal of Historical Sociology* 29 (1): 4–22.
Van der Tuin, Iris, and Rick Dolphijn. 2010. "The Transversality of New Materialism." *Women: A Cultural Review* 21 (2): 153–171.
Whyte, Kyle. 2017. "Indigenous Climate Change Studies: Indigenizing Futures, Decolonizing the Anthropocene." *English Language Notes* 55(1–2): 153–162.

8 Land Agency and the Animacy of Stories in Danis Goulet's and Amanda Strong's Short Films

Andrea Ruthven

Introduction

This chapter attends to the ways in which land is inherently agential and communicative. It draws on the work of Warren Cariou (Métis) and Vanessa Watts (Mohawk/Anishinaabe) as frameworks for considering how "it is the land itself that communicates to humans through stories" (Cariou 2020, 2). Thus understood, agency is not limited to humans or nonhuman animals, but rather the land itself is animate and "full of thought, desire, contemplation and will" (Watts 2013, 23). Epistemologies of land aliveness generate profound shifts in comprehending relationality between environments and inhabitants, questioning settler-colonial narratives of society as human action on the world. These ideas will further be considered alongside Danis Goulet's (Cree/Métis) short film *Wakening* (2013) and Amanda Strong's (Cree/Michif) stop-motion film *Biidaaban (The Dawn Comes)* (2018) to consider how their works depict settler-colonial landscapes as sites for the reclamation of Territory—in which "the deep past" and "the foreseen future" are part of "an ongoing relationship" and colonialism is a "temporary rather than [a] permanent reality" (Cariou 2020, 5).

Agency, as Jerry Lee Rosiek, Jimmy Snyder, and Scott L. Pratt (2020, 338) remind us, is something always already inherent *in* and *to* both humans and more-than-humans, and is not bestowed by humans onto other beings. Rather, the agency of humans is enabled through intra-relations with specific, particular nonhumans and the land, in which agency is mutually constituted. That is, there is no human agency without the agency of the surrounding place. There is a necessarily placed-ness, the necessarily located-ness of agency and becoming, and land aliveness is rooted (pun intended) in the necessity of being-in-common. There can thus be no presumable separation, an 'us' and 'them,' as Sean Cubitt theorizes in "Ecocritique. Against Environmentalism," as he reminds us that the word "environment" "presumes something that *environs* and something *environed*" (2019, 1; my italics). He adds that, "It seems safe to presume that the only ones talking about environing are human and that the environment is the nonhuman that surrounds them" (1). For Cubitt, we must go a step beyond the presumption

DOI: 10.4324/9781003468028-10

that environs and the environed only exist in relation to each other, and work from the logic that these are colonial categories of knowledge that divide the environment based on a human and nonhuman binary which supports rationales of dominion and mastery. We (humans) are not environed by the nonhuman (environs), but rather all forces, living and non-living, mutually constitute an environment not fractured by the false nature-culture divide. The environs and the environed as separate entities must be reconceptualized through the understanding that they are not distinct beings (human and nonhuman) but rather interdependent constructs reliant on Cartesian binaries of thought. Indeed, in order to better grasp the relational mutuality of being, I turn to consider the concept of Place-Thought, as articulated by Mohawk and Anishinaabe scholar Vanessa Watts (2013) in the article "Indigenous place-thought & agency amongst humans and nonhumans (First Woman and Sky Woman go on a European world tour!)," and that of Territory, as theorized by Warren Cariou (2020), Métis scholar, in "Terristory: Land and Language in the Indigenous Short Story—Oral and Written."

Through reading both Watts and Cariou, Place-Thought and Terristory are crucial ideas for thinking about the way in which land, seen not as an inert substance but, as Watts argues, as animate, "goes beyond being alive or acting, it is ... full of thought, desire, contemplation and will" (2013, 23). The animate, desiring and thoughtful nature of land is also asserted by Cariou who recognizes that the land communicates, stating that, by understanding Terristory we accept that "instead of humans telling stories to mimetically represent the land, it is the land itself that communicates to humans through stories!" (2020, 1–2). This recognition of land as communicative, as animate, and alert, reverts narratives of domination and oppression over the land—that is, Western, settler-colonial, capitalist narratives in which the land is an object to be exploited—arguing instead for a land that has its own agency. Settler coloniality

> has disproportionately affected many communities who fall outside of the discourse of the human—the black, the disabled, the queer, the indigenous, the colonized. Equally absent from discourses of the human are the essential forms of kinship indigenous people maintain between nonhuman animals and the earth.
>
> (Simmons 2019, 175)

By recognizing land agency, epistemologies of land aliveness contest anthropocentric narratives that reinscribe "the violence of Empire" (174) by countering stories of human exceptionalism (especially those with a narrow definition of 'human').

Thus, as Cariou argues in "Sweetgrass Stories: Listening for Animate Land":

> But when we are reading Indigenous stories, I feel it's crucial to think seriously about the ways in which the stories do, in an essential way,

> come from the land. To accept this idea is to open oneself to the possibility of an Indigenized ethic of interpretation, one that is more embodied and more active than the hermeneutic practice of reading is usually considered to be.
>
> (2018, 341)

Rather than an object that is subject to the will or desire of humans and even nonhuman others, land has its own agency, one that it exerts in conjunction with those who live with it, and can be understood through the stories it tells. Stories too are agential beings, as Eva Garoutte and Kathleen Wescott (2013) argue in "The Story is a Living Being: Companionship With Stories in Anishinaabeg Studies." Garoutte and Wescott admonish that the story is not a passive agent, awaiting representation and interpretation (especially not by white settler-scholars). Rather, researchers cannot "make choices without realizing that these will affect what their stories can *do*: those stories' abilities to evoke realities and to illuminate possible lives" (2013, 76). They highlight the necessity of an ethical reciprocity in relation to the stories we tell, the stories we dissect, analyze, and theorize with. For Rosiek et al., this "involves considering not just representational accuracy, but the relational futurities we participate in creating with other non-human agents" (2020, 342)—stories included.

Agential realism and intra-action are a mode of understanding that recognizes the ways in which our entanglements with all beings, human and nonhuman, even living and non-living, necessarily involve "risking the transformations of self and being that allegedly come with particular intra-active relations" (Barad 2007, 336). The so-called affective turn can help us understand the inherently changing and changeable selves that emerge when the Cartesian, closed self is challenged. It is specifically this risk of transformation, or, perhaps in other terms, the inevitability of transformation, if we consider the ways in which agency and being is in a constant state of reciprocal becoming, that is crucial to understandings of land agency and animacy.

Returning to Cariou, "The idea of the world around us as inherently vibrant and alive is essentially a starting point for most Indigenous ontologies" (2018, 340). The land is animate and has its own story, whether it chooses to share that story, or whether we are attuned to hear it, or not. As Kim TallBear has pointed out, "nonhumans are agential beings engaged in social relations that profoundly shape human lives" and that further to this, "for many indigenous peoples, their nonhuman others may not be understood in even critical Western frameworks as *living*. 'Objects' and 'forces' such as stones, thunder, or stars are known within our ontologies to be sentient and knowing persons" (2015, 234). Thus, understanding the nonhuman as not just living but as sentient and knowing persons situates the land and all human and more-than-human others as part of what Watts argues constitutes 'society':

> 'society' from these perspectives revolves around interactions between these worlds rather than solely interactions amongst human beings. Both of these accounts describe a theoretical understanding of the world via a physical embodiment—*Place-Thought*. Place-Thought is the non-distinctive space where place and thought were never separated because they never could or can be separated. Place-Thought is based upon the premise that the land is alive and thinking and that humans and non-humans derive agency through the extensions of these thoughts.
>
> (2013, 21)

It is not, then, agency as a human attribute that is conferred on nonhuman others in a gesture of recognition. Rather, *human* agency is dependent upon the extension of nonhuman agency and thought. Isabel Altamirano-Jiménez and Nathalie Kermoal articulate it in terms of relationality, such that "Indigenous knowledge systems have developed over millennia and are grounded in relational schemas .... These relationships are reciprocal and develop among people as well as between people and non-human beings" (2016, 7). They also point out the necessity of recognizing the local, situated character of these relationships.

Terristory, like Place-Thought, is intrinsically situated—land and stories cannot be separated but instead are mutually constitutive. For Cariou

> terristory can be understood as the ground of culture; the living, nurturing, relational medium in which Indigenous communities flourish, and also the entity or being(s) from whom Indigenous people learn their responsibilities. Terristory is a relation, or rather a plural and ongoing set of relations.
>
> (2020, 2)

This plural, ongoing set of relations, is one of mutual respect and response-ability, but one that requires the recognition of agency beyond the human, beyond the animate. Or, perhaps more clearly, it requires the recognition of the vibrancy and animation of all beings, land, rock, stones, lightning, and more. For Watts,

> habitats and ecosystems are better understood as societies from an Indigenous point of view; meaning that they have ethical structures, inter-species treaties and agreements, and further their ability to interpret, understand and implement. Non-human beings are active members of society. Not only are they active, they also directly influence how humans organize themselves into that society.
>
> (2013, 23)

How, then, can we understand the implications of habitats and ecosystems as societies? In what ways must we reorient ourselves, not just our thinking but also our material selves, in order to better engage, ethically and reciprocally, with the societies around us? Drawing on Janice Vis-Gitzel's work "The Plants Are Plotting," we must first be cognizant of how Indigenous narratives retain "a sense of land agency, a sense that the land is *living*—the place acts and is not merely a stage for human actors" (2022, 97). Although Vis-Gitzel is writing about Canadian prairie narratives specifically, I would highlight her attention to the ways in which "Ecosystems are social, comprised of the interactions of many beings who live together; this collective living is not chaotic or haphazard. Instead, ecosystems organize the flow of matter and energy, developing place-specific customs while remaining open to adaptation" (99). These interactions and social organizations are focused on the ways in which human actors are only one aspect of community, and that the plant and animal and soil others are as equally imbricated within the organizational systems. Indeed, Vis-Gitzel notes, "more-than-human communities are organizations administering energy, nutrients, and resources through systems that enable, produce, and care for relational beings living together" (99). That is, the relational and being-in-common of more-than-human communities is a foundational principle that orients all beings, human and nonhuman, living and non-living, in such a way that the distinction between them is collapsed. Through her reading of Martha Ostenso's novel *Wild Geese*, Vis-Gitzel articulates the agential power of plant politics: "The family's behaviors—the habitual movements ingrained in their bodies—are inseparable from the needs of the crops. Their plant-dependent routines shape the materialization of farm life, and the organizing power of plants becomes palpable" (100). Plant agency here is articulated through the material, corporeal needs of human agency.

The removal of nonhuman agency, according to Watts, effects the removal of nonhumans from society, and this "is accomplished in part by mythologizing Indigenous origin stories and separating out communication, treaty-making, and historical agreements that human beings held with the animal world, the sky world, the spirit world, etc." (2013, 20). The incorporation of Indigenous epistemologies into Western colonial frameworks, as a means of rearticulating our relationship to the world we live in,

> traces Indigenous people not only as epistemologically distinct but also as a gateway for non-Indigenous thinkers to re-imagine their world. In this, our stories are often distilled to simply that—words, principles, morals to imagine the world and imagine ourselves in the world. In reading stories this way, non-Indigenous people also keep control over what agency is and how it is dispersed in the hands of humans.
>
> (2013, 26)

I want then to attend to the imperative not to mythologize, not to imagine the stories of Terristory as allegory or narrative but rather as accounts, as Watts asserts, of events that took place and not as "symbols or abstractions .... They were not imagined or fantasized" (2013, 21). This imperative is especially important as the following sections think with the Weetigo and Sabe, two beings whose materiality and being are intrinsically land-based.

Wakening

Place-Thought, then, is "the network in which humans and nonhumans relate, translate and articulate their agency" (Watts 2013, 27) and Terristory is an ongoing relationship to the land, in which ground and story are linked "in a profoundly material way" (Cariou 2020, 3). This section considers how Danis Goulet's *Wakening* enacts these epistemologies. The film is an example of Indigenous slipstream narrative, what Grace Dillon defines in the following way:

> incorporating time travel, alternate realities, parallel universes and multiverses, and alternative histories is a hallmark of Native storytelling tradition, while viewing time as pasts, presents, and futures that flow together like currents in a navigable stream is central to Native epistemologies.
>
> (2012, 345)

These overlapping times, in which there is no clear division between past, present, and future, rather in which these are plural and mutually constitutive, is important for understanding how Place-Thought and Terristory function within a colonial landscape. For Dillon, Indigenous slipstream overlaps time: the past, present, and future all exist simultaneously. How we imagine the future rewrites our present and past, and how we narrate our past and present, in turn, rewrites our future.

In thinking through the apocalyptic narratives often used to discuss climate crises, Kyle Powys Whyte argues that "the hardships many non-Indigenous people dread most of the climate crisis are ones that Indigenous peoples have endured already due to different forms of colonialism: ecosystem collapse, species loss, economic crash, drastic relocation, and cultural disintegration" (2018, 226). In this way, as Salma Monani suggests, "*Wakening*'s evocation of Native slipstream—this future world, whose terror is felt in the present moment of viewing—is aimed to awaken its audiences to the colonial monstrosities that frame the horrific contexts of Indigenous lives" (2016, 13). And yet, the film does not enact "the endlessly rehearsed landscape of death and disaster that dominates contemporary visions of the coming decades" (Canavan 2014, 43). Rather than insisting on "necrofuturism" (Canavan 2014), the film invokes the potential for resistance and resurgence, to borrow from Leanne Simpson (2011), through the film's narration. As Allison Mackey suggests, Goulet's short film explores "alternative futures where humans re-establish a sense of relation with their non-human counterparts" (2018, 530).

Wakening is set in a futuristic but apocalyptic Toronto, where the Cree warrior Weesakeechak is searching for Weetigo,[1] who, as Leanne Simpson tells us, is

> a large monster-like creature that is cannibalistic [and] one of the characters in Nishnaabeg worldview that embodies imbalance and unhealthy relationships .... Wiindigo had an insatiable hunger for human flesh and was created in times of great hunger, when the temptation to eat human flesh was the greatest. ... [and] symbolizes the potentially addictive part of the human condition—when certain desires are indulged, this stimulates more indulgence until all reason and control are lost.
>
> (2011, 70)

In the film, Weesakeechak moves through the decaying remains of the city until she locates Weetigo in an abandoned theatre. Once there, she attempts to convince Weetigo to leave the cinema and move out into the world. Returning to Allison Mackey,

> Wesakechak coaxes the monster out of its lair so that it can unleash its hunger on the occupying forces instead of consuming individual human beings 'one by one.' Though the Weetigo is a voracious, cannibalistic monster, it is never totally 'othered' in Cree stories. ... Weetigo and Wesakachak are often figured as doubles of one another and, by extension, of the human/non-human relationship.
>
> (2018, 533)

Just a reminder from Watts, however, that these are not metaphorical figures. Weetigo and Weesakeechak are not narrative tropes, but rather figures of Place-Thought, in which the land communicates its being through the two protagonists.

Figure 8.1 Weesakeechak navigates the urban environment.

In the opening, we can see Weesakeechak, in dull gray tones, carrying a bow and arrow and moving quickly through an urban landscape littered and decaying, with the sound of gunfire in the background. The fear, or, as Monani terms it "futuristic ecohorror" which "produces horrific feelings in the moment of its viewing that are inevitably entangled with the past, inviting its audiences to experience the monstrous contexts of Indigenous lives across time" (2016, 1), is palpable from the opening sequence. We can hear Weesakeechak's labored breathing as she wends her way through alleys, at one point flattening herself against a wall to avoid being seen by, what we later learn must be, the occupiers. Their presence is signaled by the rhythmic, ominous sound of their boots on the pavement, and their radios squawking out unintelligible words.

As Weesakeechak surveys the urban rubble, loudspeakers broadcast a warning about illegal land ownership, an explicit gesture to the ways in which Indigenous peoples were forced off their land by settler-colonial invaders. Further, the presence of Weesakeechak, and later of Weetigo, maintains these two figures within both the narrative present and future, while gesturing to the past. Dillon's Indigenous slipstream is here made manifest as the story of colonial violence is expressly reconfigured in the future, arguing that it is not a past story, but one that pervades our understandings of the present and future. At the same time, the futurity of the film counters what Kim TallBear, among others, recognizes as "A shared narrative, that of the vanishing or dying Native, [which] has framed the response to multiple literal and figurative bodies—indigenous bodies, the land, and the indigenous body politic—by the state" (2015, 232–233). Both Weesakeechak and Weetigo, as well as the captives in the theatre, survive into the future and are revitalized through the narrative and through the dual protagonists' actions.

Weesakeechak climbs the stairs of an abandoned theatre and we hear only her breath and some bumps and creaks from somewhere off-screen. We see only what her flashlight illuminates, and the occasional glimpse of her face. The darkened space, as Monani suggests, "further disorients the viewer by making sight unreliable, forcing her [Goulet's] audience to interpret the film through sound" (2016, 5). Like Weesakeechak, we must rely on our other senses to navigate, and the scene communicates viscerally through the fear of the unknown. This darkened space, as the camera pans around, is revealed to be a now-defunct cinema. A tree and other vegetation now grow inside, and at least two humans are trapped in the seats, at first surprised, but then relieved to see Weesakeechak, believing that she has come to save them. Here, as Mackey states, Weetigo has "the power to recreate the forest, embodying the simultaneously creative and destructive potentialities of non-human agency" (2018, 533). In *Wakening*, Weetigo is powerful enough to recreate nature inside the old cinema, forging, in my reading, a critique of extractive cinema in which the natural is mined for its mediatic potentialities. Instead, the captivity of consumption through media is here rendered material, as the audience in the theatre is literally captive (replicating the captivity of the

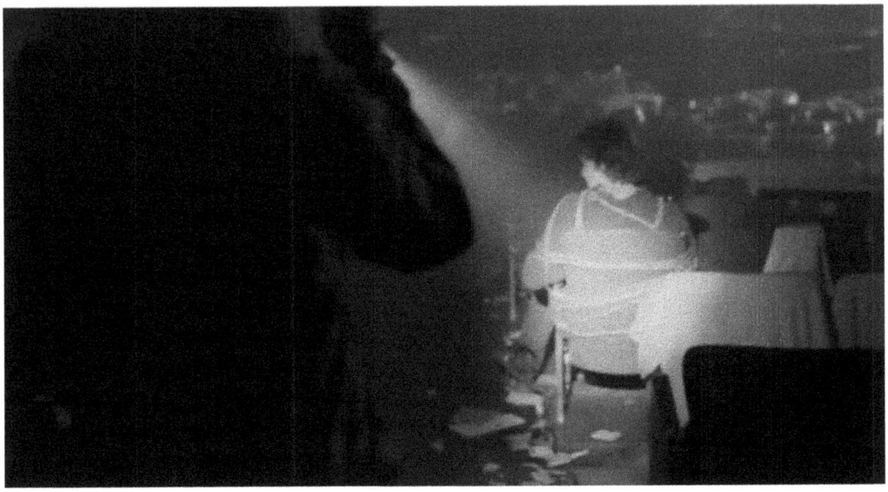

Figure 8.2 Weesakeechak surveys the captives.

viewing audience, who want to know 'what will happen'). But instead of the audience feasting their visual senses, they become the feast, though clearly Weetigo is never satiated. Rather, the audience hears the tactile crunches and moans of the previous victim; they are surrounded by the whispering dark green leaves and know only their own eventual fate.

One of the captives tells Weesakeechak: "The seats were full. We're the only ones left. You are...You are Weesakeechak." The reference to being "the only ones left" speaks not only to the captives' presence in the theatre but also to understanding Weetigo as the animate force of greed and capitalist vampirism, as the full-scale destruction and eradication of Indigenous lives from settler-colonial territories. Thinking of this scene in connection with Terristory and the importance of situatedness for agency, the viewer also sees how Weetigo's agency, located within colonial territory and trapped within colonial logics and relations, is also a colonial agency, that feeds and is fed through the intra-relations with the environment. Even Weetigo's use of Cree, as will be discussed below, is not enough to shift the power inherent in the land until they are met by Weesakeechak. It is only when Weesakeechak enters the theatre that the interrelations shift, and agencies become and re-form around, with, and through the two of them.

Viewers then witness the negotiation between Weesakeechak and Weetigo, the attempt by both to tell the story of land to the other. Monani argues that

> Significantly, the Weetigo speaks in Cree, while English is subordinated to subtitles. Weesageechak, who speaks in English, must also switch to Cree in her communications with the Weetigo. This move to privilege Cree, one might argue, is a particularly powerful decolonizing tactic.

> Following the disorienting affect of the earlier sequence—which had no words, just creaks, clanks, and unstable sight—the centering of Cree reframes the audience's sense of the locus of privilege, giving voice to language that was deliberately forbidden by colonial powers.
>
> (2016, 9)

Arguably, it also gives voice to epistemologies beyond settler-colonial logics and reinforces the agency of language and stories themselves. Neither language, in this case Cree, nor story, that of Weesakeechak and the Weetigo, are inert, both are agential and animated in this meeting between the two sentient beings. As Lisa Jackson (2018) says, the languages have been on the land for thousands of years and can thereby communicate it in a way that settler-colonial languages cannot.

The visual aspects show us the intra-action, the being-in-common of agencies, but the aural and oral aspects are equally foregrounded. As Alexa Weik von Mossner states "we need our senses in order to be able to perceive things, but also … our bodies act as sounding boards for our mental simulations of storyworlds and of characters' perceptions, emotions, and actions within those virtual worlds" (2017, 3). I would add to this that we also need our senses in order to perceive the story the land is telling us. This reliance on other perceptions is key to understanding the agency of nonhumans as the film obliges the viewer to hear the sounds of the forest, of hunter and prey, of the crunching of bone and viscera. The film also heightens tactility, the dust motes swirling about Weesakeechak, the sounds emerging from Weetigo, the tree and other greenery emerging in the darkened space, and, of course, as Monani argues, the role of language. Weetigo, as nonhuman other, has created a space which speaks to Weesakeechak and speaks to the viewers. The story told, of settler-colonial violence, speaks through the animate others that surround Weesakeechak. And when she communicates on Weetigo's terms, they are able to hear and understand each other. This too is what brings Weetigo out of the 'prison' of a metaphorical past—locked away in a now-defunct reminder of capitalist greed—and into a resurgent present. As the film ends, Weesakeechak is confronted by two occupiers, and saved by Weetigo, who has now turned their voracious hunger toward other sources.

While the urban setting of a post-apocalyptic Toronto may, at first glimpse, appear to be counter to epistemologies of land-based ethics and agency, it is worth bearing in mind Cariou's assertion that

> it is very possible to maintain a relationship of terristory within an urban context or some other overtly colonized landscape: because in a sense, terristory is more about the human attitude toward the land and stories than it is about 'what has happened to the land,' or how it has been reconfigured to suit colonial uses.
>
> (2020, 5)

The cinemascape in which Weesakeechak finds Weetigo, ostensibly freeing them from the colonial nightmare, suggests that the stories appropriated by settler-colonial forces, stories that would see Weesakeechak and Weetigo bound within discourses and narratives palatable to white audiences (as represented by the cinema as prison) are living entities and, if attended to, are still vibrant and animate beings that work against the occupiers.

I would argue, with Cariou, that Terristory is a "re-telling," it "is a way of reclaiming and recontextualizing the story, making it live again, and connecting it to … specific experiences on particular land" (2018, 69). The film emphasizes and honors the agencies of Weesakeechak and Weetigo. In this sense, Weetigo is not entirely anthropomorphized and acts according to an agency that is in constant becoming through the intra-relations with its environment. The film expresses, as Kyle Powys Whyte notes, "Weesageechak's responsibility to respect and confront Weetigo and Weetigo's responsibility not to be fooled by the occupiers. Of course, the solution to surviving the dystopia" the film indicates "lies in the reciprocal responsibility of both protagonists to work together in ways that honor each other. One way of interpreting *Wakening* is as such an unfolding narrative of dialogue with ancestors and descendants" (2018, 231), crossing or, rather, resisting the neoliberal imperative of time as necessarily forward-moving and also the segregation of generations and the privilege of youth over aging.

In the end, as Weetigo leaves the cinema and kills the two occupiers who threaten Weesageechak, there is a moment of collaboration; "*together*" as Monani suggests,

> they present alternative ways of living that thrived before Western capitalism intruded. The ending, which involves their collaborative 'going out into the world' and Weesageechak's momentary smile, is Goulet's invitation to hope, to think of how the *fantastical* of challenging occupation may be *made real.*
>
> (2016, 18)

It is this invitation to hope, that, in Terristory, "allows us to see colonialism as temporary rather than as a permanent new reality, while terristory is understood as an ongoing relationship, from the deep past to the foreseen future" (Cariou 2020, 5). Colonial logics of extraction are thus rendered objects that are part of only one story, at one moment, that need not be the defining one. The attempt to coopt and turn Weetigo to their own means, to embed within the story of the land settler-colonial logics and narratives of dominance, are here resisted and rewritten.

Biidaaban (The Dawn Comes)

Amanda Strong's *Biidaaban (The Dawn Comes)* is based on works by Leanne Simpson ("Caribou Ghosts and Untold Stories," "Plight," "The Gift is in the

Making"). Biidaaban is a genderqueer person who lives outside of a suburban neighborhood, in a one-room apartment in the basement of a house where they are visited by Sabe, an ancestral being who has a bodily materiality; they can drink tea and text with Biidaaban, but they are also depicted as transforming, materializing and vanishing, or even becoming light, throughout the film. It is important to note that the film reproduces multiple temporalities simultaneously, as the past of the Sugar Bush that Biidaaban and Sabe frequent is seen as co-existing with the present, colonized land. The future is also present, as the film opens possibilities for recognizing being and becoming beyond the logics of capitalist modernity. Further, the existence of Biidaaban and Sabe resists the aforementioned narrative of the dying or vanishing Native which are also part of what Kali Simmons identifies as the logics of assimilation which have "rendered indigenous people a figure of the past, meaning that there are no *real Indians* left, and those *real Indians* only existed in 1492, 1776, 1879, or 1934 depending on the style of settler-colonialism one invests in" (2019, 178). The state may wish to frame indigeneity as lost or past or as a nostalgia for what once was. But Strong's and Simpson's narratives ensure that indigeneity remains vibrant, living, and engaged in the world, as it has always been. The narrative further ensures a future through indigenous slipstream, by suspending "naturalized assumptions regarding Native futures, opening up a rare discursive space for alternative futuring" (Lempert 2014, 164).

The two characters, Biidaaban and Sabe engage in the practice of collecting maple syrup from trees in a suburban residential area, working together to enact decolonial practices of survivance (Vizenor et al. 2008). As Matthew Harrison Tedford notes, "Although the film seemingly represents a singular activity—urban syrup making—that activity and the struggle it entails enact a process of world building that defies the repressive logic of the settler colonial state" (2022, 46). The world building the two engage in is implicated within a politics of care and being-with that spans both physical and temporal realities that overlap and merge in an Indigenous slipstream that recognizes land agency. As we will see, the land in the film is active, agential, and becomes with Biidaaban in their attempts to engage Indigenous practices of syrup gathering. Not only this, but a non-Anthropocentric kinship opens up through the mutual practice of syrup making. As Simmons notes,

> Indigenous peoples have been offered access to the position of the human only through disavowing their kinship connections. Often, this loss of kinship connection entailed a fundamental shift in how indigenous peoples conceptualized and were able to enact kinship, particularly with non-human others such as land.
>
> (2019, 175)

Thus, through reclaiming their kinship with the trees and with each other, Sabe and Biidaaban reject colonial frameworks for the human.

The film opens with the image of a ball of blue-green light, moving through a rhizomatic space (that could be root systems or nerve systems), to become the sun and then the moon, shining over a suburban neighborhood, before reverting to the blue-green form, and traversing the streets lined with houses and maple trees. Then the image pans down through the leaves of a Sugar Maple tree, the tree that, for Robin Kimmerer (2013), contains the knowledge of the place and the land, through time. As we move through the leaves, we are confronted by the orange fencing that surrounds the base of the tree, before we see the caribou ghost of Simpson's poem (which is narrated throughout this introduction) nuzzle a patch of lichen.

This orange mesh surrounding the tree functions as a symbol of settler coloniality as it highlights the opposite of the rhizome portrayed in the opening—it is separation, distance, the individual over the collective, the protection of one against a shared being in common. Indeed, even the color, a Day-Glo orange-red that is almost the same as the bright red maple leaf, is both an identifier for the tree itself and an appropriated symbol of settler colonial domination through the Canadian flag.[2] This reminder of the insidious ubiquity of settler coloniality, the so-called 'protective' fencing, which may at first appear to have a similar rhizomatic structure to that of the opening images, is rendered a literal plastic prison, echoing the moves that the

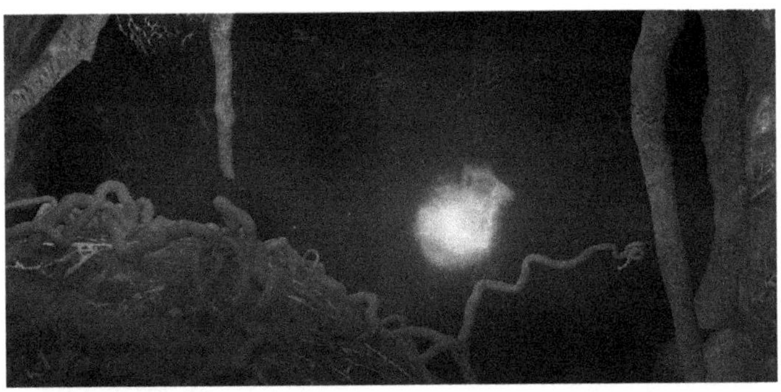

Figures 8.3 and 8.4 The radical connectivity of beings as symbolized by the rhizomatic structures.

houses and other objects of settler coloniality attempt to imprison the land and its agential inhabitants.

Through the opening sequence, the slipstream narrative insists on the agential environment, as the colonial artifact, the orange netting, exists in time with the caribou ghost, long chased from this part of the land in order to make room for the housing settlements, as well as the contemporary time of Biidaaban themself. The film continues until we are taken to Biidaaban's home, a cozy, underground space that is also inhabited, at that moment, by Sabe. Time's overlap continues as

> Strong and Simpson's depictions of Sabe do not place the character firmly in the past or the present, and they resist other binaries, such as technology versus nature. In *Biidaaban*, they are co-present with Biidaaban while also thousands of years old; Sabe is ancestral and partly composed of plant matter (and even the puppet used in the film contains real lichen), and yet they can text.
>
> (Harrison Tedford 2022, 55)

The characters communicate through a device that resembles a rock, but upon which words in Anishnaabemowin appear as the pair text each other. The land's own story and agency are here brought to the fore as the rock-like device connects Sabe, the thousand-year-old-being, with Biidaaban, in their journey to revive the practice of being with the land through maple syrup collecting. The story of the land opens to include not just the sugar maples, but also Sabe and Biidaaban, and narrates itself through their being-in-common. There is no generational or technological divide between Biidaaban and Sabe, but rather a shared relationship of care as they take turns waking each other and reminding each other of the task at hand. As with *Wakening*, language here functions as a means of communicating the past, present, and future of the land by highlighting its own agency—the agency of language—to be *with* the characters and stories told in the films.

As Sabe and Biidaaban venture out to begin their sap collection, they move into a landscape that is rife with settler-colonial imagery. The trees they seek

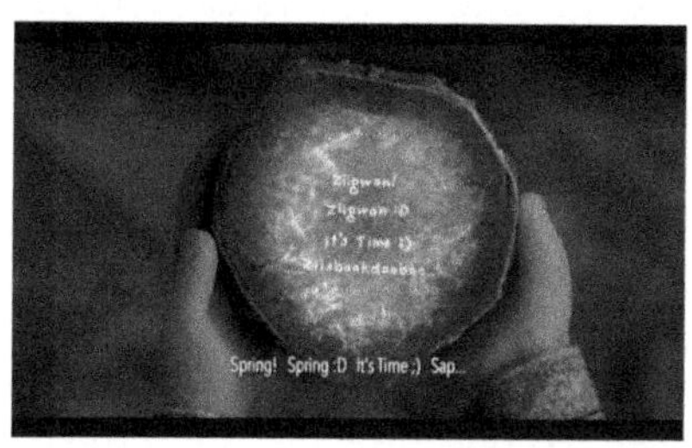

Figures 8.5 and 8.6 Sabe and Biidaaban communicate in Anishnaabemowin, combining old and new technologies.

to tap reside along the sidewalks of a subdivision, populated by two-story single-family homes. The suburban landscape looms, threatening, cut-off, as the houses seem to block Biidaaban in. The dark windows appear to be watching, monitoring a supposed invader in their territory. In another scene, the same houses appear to move, pushing in toward Biidaaban, encircling as though imprisoning. This sense of prison is further highlighted by the spotlights that come on with the hyper charged sound of a high-voltage switch. They capture Biidaaban and remind them that they are out of place. This dislocation between Biidaaban and settler-coloniality is underscored as land then becomes not a subject of agency, animacy, or mutuality, but a resource to be occupied, an inert space upon which to erect houses that keep the outdoors, and interspecies connections, at bay. As Harrison Tedford notes, "Biidaaban is treated as both a fugitive from the colonial subjectifying apparatus and an intruder on their own land" (2022, 63). The objects of coloniality—the houses, the hose for the sprinklers, the sprinkler heads themselves, all come alive in an attempt to expel Biidaaban from the space. The land itself rises up to meet them, giving them a safe place to step in their attempt to flee the house. The objects that threaten Biidaaban, the sprinklers, the hose, the lights, and the sidewalk are all tools for dominating nature, not living with it. Coloniality is here given its own agency, its own logics, as it attempts to tell a story of dominance and mastery, while the land counters with an agency located in mutuality.

As Whyte notes, there are various colonialisms and these "have ended Indigenous peoples' local relationships to thousands of plants, animals, insects, and entire ecosystems" (226). And yet, these relationships, or kinships, have not been completely obliterated, continuing as they do as a form of "felt knowledges" (Million 2009), which involve "direct ecological interaction" (Whyte 2018, 226). Through Audra Mitchell (2016), Whyte argues that the current anthropocentric discourse ignores "Indigenous peoples' experiences of having their relationships with nonhumans greatly disrupted by colonialism" (Whyte 2020, 77). In the film, we can see how this is linked to the rhizomatic, reciprocal understanding of living with the land, through the illumination of the tree as Biidaaban approaches, and Sabe melds with it, turning into light and illuminating the rhizomatic structures both above and below ground. The radical relationality (Yazzie and Baldy 2018) of Sabe, Biidaaban, and the world they inhabit moves beyond the potentially metaphorical, as the film presents as literal the animacy and agency of all beings—human and more-than-human.

Again, the plastic webbing around the trees, like the gardening implements before it, signifies the agency of colonial actions as well, competing with the agency of the trees, the ground, and Sabe and Biidaaban. The plastic fencing attempts to disrupt the sap collection by knocking the buckets Biidaaban has placed to collect the sap off of the tree. As with the rest of the film, the images here are palpable: the rough bark, the prickly, suffocating fencing, and even the metal bucket are tactile. The bucket does not, perhaps cannot, fight back

Figure 8.7 Indigenous slipstream is integrated into the closing sequence.

against the violent creep of the fencing, and the insidious spread of the plastic webbing up the tree attempts to negate the reciprocal living with that the bucket and sap collection represent. However, this same living-with is victorious, as Biidaaban pulls back the plastic and sets the bucket to rights: The bucket, the sap that drips into it, and the ways in which these work together tell the stories of the land, both prior to and during settler-colonialism, but also into the future. Indeed,

> Biidaaban and Sabe's sap-collecting practice [is] a world-building project that is much more than a dream. Given the centrality of sugar maples to people of what Kimmerer calls 'Maple Nation,' Biidaaban serves as an embodiment of resistance and as a way of creating an anti-colonial, anti-capitalist world.
>
> (Harrison Tedford 2022, 50)

The collection of sugar maple sap and the process of making syrup take on deeper meaning as they are indicative of the survivance, resurgence, and slipstream of Indigenous art production that are in direct contrast to the extractivist white settler-colonial maple sugar production processes. The world being built through Biidaaban and Sabe's 'embodiment of resistance' is tied to the stories the land has to tell and visualized in the final image of the film, in which the three temporalities that modernity would sunder—past, present, and future—are joined within a circle of mutual being and becoming.

Conclusion

Following the warrior Weesakeechak's attempts to locate the Weetigo, *Wakening* illustrates the way in which the land's story is not separate from that of the humans and nonhumans that populate it, but rather is agential and animate through its connections. Instead, the film suggests that land agency, and the ability to communicate with the land and have it choose to communicate as well, may be muted within settler-colonial habitats, but are nonetheless still present. *Biidaaban (The Dawn Comes)* also enacts resistance to settler-coloniality's attempt at mastery through territorial and epistemological frameworks. Through Biidaaban and Sabe's practice of being with the land through the collection of maple syrup, the agency and animacy of the area—the land, plants, air, and more-than-human others—are made literal, belying both the metaphorical and symbolic destruction of Indigenous peoples and practices. As Blair Topash Caldwell mentions, "Indigenous experiences with issues attributed to the Anthropocene are different; Indigenous communities have already undergone loss of place, access to resources, species extinction, and dire social unrest because of colonialism" (2020, 85). In both films, the questions of anthropocentric apocalypse remind viewers that not all humans are anthropos, that settler-colonial practices are only one way of being in and being with the world, and that Indigenous futuring is both possible and present.

For Harrison Tedford, "Certain works of Indigenous futurism, such as *Biidaaban*, are capable of resisting the illusion of colonial permanence because of their common acknowledgement of already existing dystopian conditions, prioritization of direct embodied action, and reliance on inter-generational thinking" (2022, 51). The privilege of youth and the rejection of the living past brought about by settler-colonial dreams of neoliberal individualist culture are countered by the intergenerational relationships and knowledge sharing engaged in through the maple syrup making process and Sabe and Biidaaban's relationship. The same could be said of *Wakening*, as it engages Indigenous slipstream as a means of resisting colonial narratives of a vanishing or obsolete past linked to linear temporalities. Through Terristory, these films are a way of narrating "how to survive on the land" (Cariou 2020, 3). For both films, the animacy and agency of the land, as dramatized through Terristory and its ability to story its own being and engage connections, through language, with its human and more-than-human others, serve as reminders of the potentiality of radical relationality and being-in-common.

Notes

1 Spelling for Weetigo, also Wiindigo, and Weesakeechak varies depending on the language and culture being spoken, and I have reproduced the names as they are written by the different authors.
2 I would like to thank Belén Martín-Lucas for pointing this out to me.

Works Cited

Altamirano-Jiménez, Isabel, and Nathalie Kermoal. 2016. *Living on the Land: Indigenous Women's Understanding of Place*. Athabasca: Athabasca University Press.

Barad, Karen. 2007. *Meeting the Universe Halfway: Quantum Physics and the Entanglement of Matter and Meaning*. Durham: Duke University Press.

Canavan, Gerry. 2014. "'If the Engine Ever Stops, We'd All Die': *Snowpiercer* and Necrofuturism." *English Faculty Research and Publications* 298: 1–26. https://epublications.marquette.edu/english_fac/298.

Cariou, Warren. 2018. "Sweetgrass Stories: Listening for Animate Land." *The Cambridge Journal of Postcolonial Literary Inquiry* 53: 338–352. https://doi.org/10.1017/pli.2018.10.
Cariou, Warren. 2020. "Terristory: land and language in the Indigenous short Story—Oral and Written." *Commonwealth Essays and Studies* 42 (2): 1–9.
Cubitt, Sean. 2019. "Ecocritique. Against Environmentalism." *Media+Environment* 1 (1). https://doi.org/10.1525/001c.10784.
Dillon, Grace L. 2012. "Imagining Indigenous Futurisms." In *Walking the Clouds: An Anthology of Indigenous Science Fiction*, edited by Grace Dillon, 1–14. Tucson: University of Arizona Press.
Garroutte, Eva, and Kathleen Dolores Westcott. 2013. "'The story is a living being': Companionship with stories in Anishnaabe Studies." In *Centering Anishinaabeg Studies: Understanding the World Through Stories*, edited by Jill Doerfler, Heidi Kiiwetinepinesiik Stark and Niigonwedom James Sinclair, 61–88. East Lansing: Michigan State University Press.
Harrison Tedford, Matthew. 2022. "Is a Non-Capitalist Future Imaginable?: Embodied Practices and Slipstream Potentials in Amanda Strong's *Biidaaban.*" *Feminist Media Histories* 8 (1): 46–71. https://doi.org/10.1525/fmh.2022.8.1.46.
Jackson, Lisa. 2018. *Interview* "Imagine if Toronto were reclaimed by nature" *CBC Arts*. Biidaaban: First Light VR. https://www.youtube.com/watch?v=B6EUAEw1-ik.
Kimmerer, Robin Wall. 2013. *Braiding Sweetgrass: Indigenous Wisdom, Scientific Knowledge and the Teachings of Plants*. Toronto: Milkweed Editions.
Lempert, William. 2014. "Decolonizing Encounters of the Third Kind: Alternative Futuring in Native Science Fiction Film." *Visual Anthropology Review* 30 (2): 164–176.
Mackey, Allison. 2018. "Guilty Speculations: The Affective Climate of Global Anthropocene Fictions." *Science Fiction Studies* 45 (3): 530–544. https://doi.org/10.5621/sciefictstud.45.3.0530.
Million, Dian. 2009. "Felt Theory: An Indigenous Feminist Approach to Affect and History." *Wicazo Sa Review* 24 (2): 53–76. https://doi.org/10.1353/wic.0.0043.
Mitchell, Audra. 2016. "Extinction." In *Gender: Nature*, edited by Iris Van Der Tuin, 183–198. New York: MacMillan.
Monani, Salma. 2016. "Feeling and Healing Eco-social Catastrophes: In the 'Horrific' Slipstream of Danis Goulet's *Wakening.*" *Paradoxa* 28: 192–213.
Rosiek, Jerry Lee, Jimmy Snyder, and Scott L. Pratt. 2020. "The New Materialisms and Indigenous Theories of Non-Human Agency: Making the Case for Respectful Anti-Colonial Engagement." *Qualitative Inquiry* 26 (3–4): 331–346. https://doi.org/10.1177/1077800419830135.
Simmons, Kali. 2019. "Reorientations; or, an Indigenous feminist reflection on the Anthropocene." *JCMS: Journal of Cinema and Media Studies* 58 (2): 174–179.
Simpson, Leanne Betasamosake. 2011. *Dancing on Our Turtle's Back: Stories of Nishnaabeg Re-Creation, Resurgence, and a New Emergence*. Winnipeg: ARP Books.
TallBear, Kim. 2015. "Dossier: Theorizing Queer Inhumanisms: An Indigenous Reflection on Working Beyond the Human/Not Human." *GLQ: A Journal of Lesbian and Gay Studies* 21 (2–3): 230–235.
Topash-Caldwell, Blair. 2020. "'Beam us up, Bgwëthnènè!' Indigenizing science (fiction)." *AlterNative: An International Journal of Indigenous Peoples* 16 (2): 81–89. https://doi.org/10.1177/1177180120917479.
Vis-Gitzel, Janice. 2022. "The Plants are Plotting: Political Orders in Ostenso's *Wild Geese.*" *Canada & Beyond* 11: 95–113.
Vizenor, Gerald, ed. 2008. *Survivance. Narratives of Native Presence*. Lincoln: University of Nebraska Press.

Watts, Vanessa. 2013. "Indigenous Place-Thought & Agency Amongst Humans and Non-humans (First Woman and Sky Woman Go on a European World Tour!)." *Decolonization: Indigeneity, Education & Society* 2 (1): 20–34.
Weik von Mossner, Alexa. 2017. *Affective Ecologies: Empathy, Emotion, and Environmental Narrative*. Columbus: Ohio State University Press.
Whyte, Kyle Powys. 2018. "Indigenous science (fiction) for the Anthropocene: Ancestral dystopias and fantasies of climate change crises." *Environment and Planning E: Nature and Space* 1 (1–2): 224–242.
Whyte, Kyle Powys. 2020. "Indigenous realism and climate change." In *Climate Realism. The Aesthetics of Weather and Atmosphere in the Anthropocene*, edited by Lynn Badia, Marija Cetinić, and Jeff Diamanti, 69–81. London: Routledge.
Yazzie, Melanie K., and Cutcha Risling Baldy. 2018. "Introduction: Indigenous peoples and the politics of water." *Decolonization: Indigeneity, Education & Society* 7 (1): 1–18.

9 New Animism and Shamanic Cinema

Human–Animal–Machine Interactions

Marta Segarra

In the last two decades, there has been a revival of the notion of "animism"—including "animist cinema," as we shall see—stripped of the derogatory meanings that Edward B. Tylor, who launched the term in his book *Primitive Culture* (1871), had given to it (Bird-David 1999). Tylor, and the vast majority of anthropologists until the 1950s at least, saw animism as an *infantile* stage in the supposed progress of human culture, proper to "primitive," *savage* people. The term was dropped as anthropology evolved to a less western-centered discipline, but has been revived since the beginning of the 21st century under the influence of scholars such as Philippe Descola (2005), Bruno Latour (1997), and Eduardo Viveiros de Castro (2009). As is well known, animism is an ontology, prevalent in many cultures around the world, which assumes that all forms of life share a soul, rather than holding that this is exclusive to humans. Of course, in animism the expression "forms of life" also adopts another meaning than in the Global North, since some "artifacts" we might deem inanimate have an "ambiguous ontology" (Viveiros de Castro 2019, 29), as they may also, in some cases, possess a soul.

Ever since a "New Animism" (Bird-David et al. 1999; Harvey 2005; Lamarre 2008) has pervaded the cultural regions where, however, human exceptionalism still prevails, the term is often used, in anthropology and beyond, with a positive connotation. An animistic perspective of the world is even being called for in order to counter the effects of the Anthropocene or the Capitalocene, especially the global climate crisis and the extinction of so many species. Cinema is part of this turn, and some films are deemed "animist." For instance, Isaac Marrero-Guillamón (2018, 26) applies the term to Thai filmmaker Apichatpong Weerasethakul, one of whose films will be discussed here.

Since the term "soul" (or its equivalents in Romance languages, derivatives of Latin *anima*) carries strong religious, or at least spiritual, connotations, many of these new animisms prefer to speak of "agency" or even "personhood," when referring to nonhuman beings. In the wake of animal rights' activism and philosophy, the personhood of (some) nonhuman animals has been claimed since the 1970s (Cavalieri and Singer 1993). However, these thinkers advocate for a limited inclusivity: that of some mammals,

DOI: 10.4324/9781003468028-11

especially great apes, the closest species to *Homo sapiens*. Moreover, proponents of new animisms often claim that animism is a "natural" tendency of all human beings, as can be seen in young children, who tend to not discriminate between human and nonhuman animals and who often address objects as animate beings with their own will. This universalization of animism, though, is also a way of downplaying the specific features of actual animist ontologies as well as of appropriating these non-western perspectives. The association between childhood and other-than-western worldviews also points to an implicit hierarchy between adult reason and other modes of thinking.

Moreover, the generalization of personhood in animism does not imply that different categories of being-in-the-world should not be kept separate, as Viveiros de Castro states (2019, 29). When they are muddled, this is a dangerous situation for the humans involved. Shamans, as "transspecific beings" (23), are the exception to this rule, the only humans capable of engaging in "transspecific dialogues" (26). Shamanism, thus, can be defined as "the capacity evinced by some individuals to cross ontological boundaries deliberately and adopt the perspective of nonhuman subjectivities in order to administer the relations between humans and nonhumans" (26).

I will discuss in this essay what has been called "animist cinema," on the one hand, and "shamanic cinema," on the other, through several films: Apichatpong Weerasethakul's *Sat pralaat!* (*Tropical Malady* for its international distribution; Germany/France/Italy/Thailand, 2004), Francisco Huichaqueo's *Gente pájaro. Che Uñum* (Chile, 2007), James Marsh's *Project Nim* (United Kingdom/United States, 2011), and Gerard Ortín Castellví's *Agrilogistics* (Spain/United Kingdom, 2022)—although I will not devote equal attention to each of them. These films, and their authors, are diverse in terms of cultural context, technical means, year of production/age, genre, budget, promotion, and reception. For the purposes of this chapter, what is most relevant is that Apichatpong[1] and Huichaqueo have a life experience in non-western ontologies (the first was born and brought up in Thailand, and the second is of Mapuche descent), whereas Marsh and Ortín Castellví were raised and live in Europe. Regarding genre, *Gente pájaro* and *Agrilogistics* are experimental films, the first close to surrealist videoart and the second to documentary, and *Project Nim* is a more traditional documentary, but also a biopic, as I will claim. As for *Tropical Malady*, the best-known and critically acclaimed of the four (it was released at the 2004 Cannes Festival and won the Jury Prize), it has been said to belong to "realist" cinema, but not in a classical sense: Ingawanij (2013) characterizes its genre as "performative realist cinema," Quandt (2009, 66) describes it as "hallucinatory realism," Teh (2011) inscribes it in "social surrealism" and Marrero-Guillamón in "ethnographic" and "world-making" cinema (2018, 17).

Despite their differences, these films maintain, thus, cross-relationships with the documentary genre and with a twisted realism, or "alter-realism," in Martín Savransky's terms, referring to "realism that takes the risk of asserting the reality of what is deemed improbable, implausible, marginalized, suppressed, irrelevant, even scandalous, and seeks to draw out its possible implications for the transformation of what is considered credible, reliable and serious" (quoted in Marrero-Guillamón 2018, 29). I will begin with the less "alter-," or the more "realist" of the four films here discussed, James Marsh's *Project Nim*, which clearly inscribes itself in the documentary genre.

Dreams of Hybridity: *Project Nim*

Humans have always dreamed of human/nonhuman hybridization. In the Global North, this dream matches with the belief in human exceptionality, which is still largely pervasive. The term *hybrid* comes from the Latin *hybrida*, which designates the offspring of a sow and a wild boar, but it will also be significant for my argument that this Latin word is related to the Greek term *hybris* (more commonly spelled *hubris* in English), namely the excessive pride of humans who dare to compare themselves to the gods, which leads them to become monsters, according to classical tragedy (Segarra 2022). If hybridity is represented as something abnormal, it is because the "anthropological machine," in Giorgio Agamben's (2002) words, uses these mixed figures in order to reaffirm the boundaries of humanity, its categorical difference from the rest of living beings, and not to illustrate the porosity of these limits.

In fact, the advocates of racism claimed the hybrid origin of certain "races" to justify their inferiority to the "white race," with the effect of excluding these people from humanity. In the mid-18th century, Hoppius argued that some populations such as the "Moors," the "Pygmies" and "other Negroes" (as he put it) were the product of hybridization between man and ape (Newton 2011). These ideas formed the basis of the colonial ideology that flourished in the following century. In the same period, the European Enlightenment, Lord Monboddo argued, on the contrary, that anthropomorphic animals (nonhuman great apes) were capable of learning human language and that, therefore, they should be integrated into human kind, anticipating Singer and Cavalieri's Great Ape Project by two centuries.

Although there is no agreed-upon definition of "language," the mastery of language has been considered, at least since Aristotle, one of the most definite frontiers between mankind and animals. Diderot explains that the Cardinal of Polignac, contemplating an ape in the royal menagerie of Versailles and admiring their resemblance to human beings, went up to them and exclaimed: "Speak and I will christen you" (Fontenay 1998, 332). The human attempt to communicate with other species seeks mainly to benefit humans, but also arises from the desire to know—which also means to master or *domesticate*—the radically different other, what I have called "dreams of hybridity." Attempts to teach the human language to nonhuman primates, which had begun at the end of the 19th century, intensified in the second half of the 20th century (Herzfeld 2012). It was thought that it would be better for the apes to learn human language like infants do, growing up next to adult humans in a "family."

In the mid-seventies, Herbert Terrace, a psychology professor at Columbia University, started a research project with a chimpanzee he named Nim Chimpsky, with the aim of sharing language with an animal, by teaching him human (American) sign language. Nim, born in captivity, was separated from his mother shortly after birth and raised by a former student of Terrace, Stephanie, as another of her own children (she even breastfed him). After his early death, Nim was the subject of a biopic by British documentary author and filmmaker James Marsh, *Project Nim*. *Project Nim* follows a traditional biography structure: it depicts Nim's origins, his birth, the rise of the hero, his fall from grace (due to the professor's *hybris* or perhaps of Nim's, trying to trespass into human species' reserved area) and his partial redemption interrupted by an untimely death. Marsh combines the use of real footage with the reenactment of past events by actors, as well as interviews with people who interacted with Nim. These monologues (since the interviewer's questions or comments have been cut off) take place on a stark set and are presented in static, medium shots. They convey, thus, the impression of confessional speeches, as James Marsh intended (2011). And the most common feeling shared by the interviewed is indeed guilt, which can be considered a judeo-Christian frame of mind.

The documentary, especially at its beginning, which covers Nim's childhood, includes plenty of footage from Super 8 real home movies that Stephanie's family made, as so many families in the seventies. This results in a sort of imitation—which might be deemed parodic—of the American family dream. For instance, when Nim is taken from Stephanie's home, Nim, Terrace, and a young assistant, Laura, are shown in a fancy car, as if they were "dad," "mom," and the only, spoiled child, driving to a beautiful mansion with a large garden. In fact, Nim is always under the care of successive couples. All these "families" are deeply dysfunctional (as perhaps all families are). Laura, who stands for the evil stepmother trying to replace Stephanie in Nim's heart, concludes: "You can't give human nurturing to an animal that can kill you," even though all human animals can kill you, including your own children when they grow up, as Oedipus' tragedy famously shows.

This inclusion of a nonhuman animal in human normativity also touches gender roles, which are central in the film: the "alpha males"—although current ethology contests this notion—are constantly challenged by other males who want to take their place (notably Herbert Terrace, who is the "villain" of the film in a certain way), but there is also a strong rivalry between the two "mothers," the "bad" mother and the "good" mother—even though both are two sides of the same coin in the child's unconscious, following a psychoanalytic interpretation. Nim's behavior is often explained in gendered and also in speciesist terms, such as: "This is the life he was hardwired for," alluding to challenging the alpha male. Implicit in this statement is the idea that animals are only guided by "instinct," unlike "rational" humans. But Nim is also anthropomorphized, as a violent young male (another female teacher compares him with a "bad boyfriend") who abuses weak women, but respects strong females who hit him back. He feels betrayed by his human mother, who also arouses him, and he is forcefully jealous of his father, in a typical oedipal frame, as described by Freud. In this sense, Nim fulfills the oedipal structure that, according to Deleuze and Guattari (1980), always characterizes the relationship between human "masters" and their "pets."

Also regarding gender, true friendship is only assumed to be possible between two males, who play silly physical games together and go out for beers, sharing a joint. Bob, a student who joins the care team, wishes to become "Nim's friend" by "building a relationship," which is not language-based, with him. Bob specifies: "I didn't care about the language. There was communication going on." James Marsh (2011) argues that in his short life, Nim encountered only one person who "trie[d] to meet [him] half-way," and who was willing thus to have *an animal relationship with the animal*, to paraphrase Gilles Deleuze (1996). Bob is the only human who respects the "commitment" they all made to Nim; all the rest "failed," as Stephanie bitterly admits. A striking scene shows this idea, near the end of the film: when Bob, after years of unsuccessful attempts, is allowed to visit Nim in the refuge for abused animals where he is living, they soon reconnect and resume their friendship. A still image of Nim's fingers sticking out of the bars in order to touch Bob's symbolizes this connection.

However, cell bars, a reiterative image in the film, stand for the ineffaceable limit between the two worlds, deemed human and animal. Nim looks often into the eye of the camera—in the scenes with human actors wearing ape-like masks, since this anthropomorphic behavior is not typical of a chimpanzee—as do the other (human) characters, but in a more poignant way, as if wanting to communicate what he cannot tell by speaking. In fact, most of Nim's life is spent inside a cage: he is taken out of it when he is ripped from his chimpanzee mother's arms, and will only live outside until Terrace brings him back, as a discarded toy, to his original owners, who sell him to a laboratory. Nevertheless, even in his short time out of the cage, Nim is kept on a leash when walking with humans from the moment he reaches puberty. As an adult, he will never be shown on the other side of the bars.

I will draw two tentative conclusions from *Project Nim*. First, this film is a biopic in a traditional way, which follows the structure of a classical tragedy, as mentioned, but focuses on a nonhuman subject. Should this be considered a typical anthropocentric view on nonhuman animals, caged in an oedipal frame, as Deleuze and Guattari (1980) denounced, or rather as a subversion of the genre that challenges the stable borders between the human and the nonhuman, since only humans are entitled to be subjects of a biography? The latter would make the film somewhat a dream of hybridity.

The second conclusion relates to language: *Project Nim* gives the impression that words are useless and even deceitful when it comes to affects, to drives, to love, and even to friendship. When Nim arrives in his first "family," one of the human children realizes that "it's alive, it's not a toy" but a "needy creature" who tries to "mold" his human siblings as well as they are trying to cast him into a human linguistic mold. His loving relationship with his first human mother, Stephanie, thrives without needing human language, according to her (and the real Super 8 footage of that time seems to confirm that). After Nim has learned some sign language, she concludes: "He was less with language as he was as its unique self"[sic]. In *Project Nim*, the humans do all the talking, but, with only one exception, they fail in being *response-able* to the demands of a relationship, as Haraway (2003) puts it. In other words, none of the humans in the film engages in a "transspecific dialogue," especially those who try to make Nim learn human language. Even Nim's and Bob's "friendship" is doomed to failure, and Bob will only be able to alleviate Nim's captivity by visiting him from time to time. Therefore, *Project Nim* may be rated a *humane* film, as a claim for nonhuman animal rights and an indictment of humans' tendency to objectify other living beings, including other people, but at the same time it stresses the boundaries between the human and its other, *the* nonhuman animal, taken as a whole. Moreover, the film's insistence on guilt—and, therefore, on atonement and redemption for the sin of excessive pride or *hybris*—sets it in an implicit judeo-Christian framework.

Shamanic Cinema: *Tropical Malady* and *Gente Pájaro. Che Uñüm*

Apichatpong Weerasethakul's *Tropical Malady* has been specifically addressed as an "animist film," and also as "shamanic cinema" (Mandelbaum 2004). Nevertheless, some critics have warned against the tendency to forget that Apichatpong's films also reflect the socio-political reality of contemporary Thailand and the effects of its recent troubled history on the population (Teh 2011). However, to consider both features mutually exclusive would be tantamount to thinking that animism is a spiritual, ahistorical—or outdated—belief that has no place in the "modern" world, whereas animism is an ontology as valid as "naturalism," in Descola's (2005) terms, despite what most people in the Global North may still think. As today's Thai culture and, above all, Apichatpong's eclectic sources show, an animist perspective can perfectly coexist with a shrewd political indictment.

That is why, although *Tropical Malady* is structured in two distinct parts, it would be misleading to consider them symmetrically opposed, the first a realist narrative clearly located in contemporary Thailand, and the second an oneiric fantasy with a tenuous connection to the former. The first part depicts the beginning of a homoerotic relationship between Keng, a soldier stationed in a remote part of Thailand, and a young worker, Tong, who lives with his family in the forest. After Tong's sudden disappearance in the dark, the film's second part, entitled "A spirit's path," takes place in the nighttime and shows "Keng's physical dive into the blind universe of the jungle" (Mandelbaum 2004). This second part has a dreamlike and even surreal touch, since the viewer does not know what Keng is chasing in the jungle or who is chasing him. There is a story about a tiger, not a real animal, but a hybrid creature, the outcome of a "metamorphosis" or a "possession," as the tiger bears the spirit of a dead shaman.

If the first part of *Tropical Malady* is more alike to Thai romance movies, including a musical scene with a pop singer, the second part embraces the atmosphere of Thai ghost tales, which Apichatpong is fond of (Ingawanij 2013). The filmmaker says in an interview (Weerasethakul 2016) that as a child he was fascinated by science-fiction novels and movies, as well as by traditional Thai ghost tales, two apparently very different genres. He mentions Philip K. Dick and Spielberg's *E.T.* as influences, but also Abbas Kiarostami, French *cinéma d'auteur*, citing Robert Bresson, and avant-garde cinema by Maya Deren and Andy Wharhol, among others. Although Apichatpong's feature films—he is also a well-known author of videoart—are more akin to the narrative fictional cinema, they also embrace experimental documentary as a genre committed to blurring the boundaries between "truth and fiction, reality and art," rejecting though the traditional documentary's claim of telling "the truth" (Teh 2011, 597). In contrast, *Project Nim*, which purports a "message," albeit complex and contradictory, as is inherent to family and love relationships, is a much more conventional documentary, for instance, in its extensive use of human language.

As for *Tropical Malady*, in its first part, Keng is constantly chatting with the aim of seducing Tong, while in the second one, the soundtrack mainly consists of forest sounds. Keng's radio only produces white noise, which seems like an unintelligible, foreign language. From then on, words are useless, although a talking monkey tells Keng that the tiger desires him both as prey and as a companion. Jacques Mandelbaum (2004) infers that this second part "reedit[s] the first story on the level of the unconscious." It could also be argued that this second part is Keng's dream, which would amount to the same. And Mandelbaum appropriately quotes Georges Bataille, who said: "Desire is to time what the tiger is to space."[2] If this second part reenacts the story of a seduction, it does so by revealing how activity and passivity, in desire, do not constitute a pair of opposites, but are rather combined in each of the two partners, who are each hunter and prey. This is as evident in the first part as in the second.

Near the end of the film, Keng ventures into the forest, walking on all fours like the tiger. The soldier's movements become less and less human-like. There is a frequent use of subjective camera, at the level of the tiger's eyes. We hear a shot and the tiger "absorbs" Keng's soul and memories. The last images are of treetops amidst the wind with a soundtrack of nocturnal animals' cries. The hybridization of the human and the tiger, but also the entire forest, is achieved, which is indeed an animist conclusion.

If critics spoke about "shamanic cinema" regarding *Tropical Malady*, its author explained that the nighttime part is about "emotion" and "out of body experience," adding that he felt a strong identification with the character walking in the forest, which he qualified as a "shamanic feeling" (Weerasethakul 2016). Moreover, he felt as if this second part were "autonomous," as if it had been filmed by the film itself, concluding: "A movie is like an animal," since "it has life." This might be correlated with Hélène Cixous's (2005) conception of writing (in the Derridean extended meaning of the term): for there to be writing, the author has to relinquish the steering wheel of the car, that is, their own mastery, to leave it to language itself, opening the way to the unconscious. This parallel with Cixous, apparently somewhat arbitrary, is relevant here, not only because Apichatpong became familiar with deconstruction during his architecture studies in the 1990s, prior to his film studies, but also because, in my view, animist ontology is kin with the European-born notion of the unconscious, as I shall develop below.

I will briefly bring up another film that could belong to "shamanic cinema" as well. *Gente pájaro. Che Uñüm* is an experimental video, of the same length as Ortín Castellví's *Agrilogistics*, authored by Mapuche scholar and filmmaker Francisco Huichaqueo. *Gente pájaro* delves into Mapuche mythology in order to reflect the alienation suffered by indigenous people forced to emigrate from their lands to Santiago, the capital of Chile. It combines, thus, in a harsher way than *Tropical Malady*, a political stance on national politics with a non-western, animist standpoint on "reality."

Gente pájaro bears eerie resemblances to *Tropical Malady*, but also to *Agrilogistics*, including the performance of a traditional song in Mapudungun, and the highlighting of a contemporary anarchist song in Spanish. Songs take a central place in these three films, whose soundtracks consist mainly of natural sounds. *Gente pájaro*'s first scene shows a man jumping from tree to tree, and another person wearing a birdlike mask and crawling on the ground. The film includes many nighttime scenes, linking images in a seemingly free dreamlike association. It may also be said that *Gente pájaro* wonders *what it is like to be a bird*, like Thomas Nagel's famous article, "What is it like to be a bat?" (1974), thus linking the film to other-than-western epistemologies (Díaz Silva 2009). In a dialogue with Huichaqueo, curator and art historian Carolina Castro (2021) suggests that "It is interesting to think of dream as a place where it is possible to preserve tradition and dialogue with the ancestors," to which the filmmaker answers: "The world of dreams, the *pewma*, is undoubtedly a space that the colonizers could not access." He adds that his intention as a filmmaker and also curator is to "resignify" cinema and the institution of the museum. Even more interestingly, Castro (2021) concludes about *Gente pájaro*: "There is something here that connects the spiritual with the technological, and opens up the possibility that a device such as a screen is also a gateway." Linking the animist—although not in a spiritual sense—with the technological is precisely what Ortín Castellví's *Agrilogistics* achieves, in my view.

Human–Animal–Machine Interactions: *Agrilogistics*

Gerard Ortín Castellví's experimental documentary, *Agrilogistics*, is also structured in two well-differentiated and symmetrical parts, the first taking place in a high-tech facility producing vegetables, fruits, flowers, and potted plants, during the working day. The second part, shot at night, is set in a similar location, but as if the greenhouse had fallen into ruins and was invaded by an abundant proliferation of weeds. When I first saw *Agrilogistics*, its two-part structure immediately reminded me of *Tropical Malady*, but I thought it was simply a coincidence. However, Gerard Ortín (in a private communication) later confirmed that Apichatpong's film had been one of his references when creating *Agrilogistics*.

The film's first part, thus, in day- and worktime, focuses on the assembly lines, which are mainly automated and conducted by industrial robots. The presence of humans in the film is restricted to small distant figures, and also to gloved hands, separated from their human owners, doing either the same work as robots or else cleaning them. The effect of this stylistic choice is a feeling of indistinguishability between the human and the machine. The use of sparse dolly shots (a characteristic trait of *Project Nim*, in which these moving shots show Nim's motion, and in a certain way, mimic or embrace it), following the pace of the assembly line, adds to this hybridization. The soundtrack consists mainly of machine noises and of a song played on a distant radio, which will be resumed more clearly at the end of the film. This first part may be interpreted as a critical view of post-Fordist dehumanization of labor and, specifically, of the manufactured treatment of "natural" products, such as vegetables and flowers, for human consumption. This is not a misreading, but, as in *Tropical Malady*, if both parts of *Agrilogistics* are considered together, the meanings that the film conveys are subtler and more complex.

The second part of *Agrilogistics* is shot during nighttime, and the greenhouse appears in a completely different state. It is invaded by an overgrown, luxurious vegetation, abandoned by the hand of humans. Suddenly, the silhouette of a llama comes into view, in a reddish light, and the figures of other nonhuman animals like sheep can also be distinguished. We hear the sound of hoof beats, a calf softly mooing, and the flapping of a bird trying to get out of the greenhouse through the glass roof. The llama looks into the camera eye, thus to the film viewers, with the seemingly indifferent, enigmatic gaze of some nonhuman animals. In contrast to Nim's anthropomorphic stare into the camera, the llama's look transmits a nonhuman alterity. This would seem contrary to animist ontology, which claims that all beings share the same essence, whether named soul or mind, although embodied in different forms. But, as Viveiros de Castro reminds us, animism is not a relativistic thought; for animists, "Perspectives should be kept separate" (2019, 29). Only some beings, the shamans, are able to cross the boundaries without losing their life; more properly, as said above, shamans are in charge of the communication between different forms of being.

Agrilogistics' second part includes a powerful, iconic scene in which the arm of a machine picks plant buds on the production line, while a long-legged insect walks on it toward the vegetation. The mechanical arm moves in a seemingly free and random way, but also gives the impression that it may be commanded by the animal.

The end returns to daytime, and a last still image shows bits of dead plants and flowers discarded on a table, inside the facility, while the song "Prayer in C" by Lilly Wood & the Prick sounds. Its lyrics allude to the end of the human world, or the end of the Anthropocene: "... our world is slowly dying," "... when seas will cover lands / And when men will be no more." Like in *Tropical Malady* and *Hombre pájaro*, a song takes the place of the dialogue; the fact that we may not understand the words sung, as well as the predominant melody, undermine the specifically human linguistic basis of these scenes.

Significantly, the second part of *Agrilogistics*, as the second half of *Tropical Malady*, has a dreamlike quality—even though the first one, under its apparent industrial, matter-of-fact objects, also has a hypnotic quality, with its static shots of the slow ballet of robots. Gerard Ortín Castellví explains, about his own filmmaking, that "the use of dreamwork and fabulation [may be used] as a counter-strategy to challenge the prevailing imaginaries of precision agriculture, enabling unexpected life encounters and proposing new ecological and technological entanglements" (2022). This aim can be related to Huichaqueo's purpose to "re-signify" cinema while decolonizing this western artifact. *Tropical Malady*, *Gente pájaro* and *Agrilogistics* thus make compatible a critical stance on current "social reality" with a foray into a dream or an animist universe, in what D. Teh considers "a neglected *relation* between the unconscious and social reality" (2011, 604).

In this sense, *Agrilogistics* might also be regarded as "shamanic cinema," escaping an exclusive western, modern perspective of the world, and thus contributing to "the decolonizing of thought" that Viveiros de Castro (2009, 3) relates to the scholarly attention given to other-than-western ontologies such as animism. Both *Tropical Malady*'s and *Gente pájaro*'s authors are familiarized with non-western ontologies but also with the European avant-garde film archive, which they combine in an original and rich manner. And this avant-garde tradition makes an intensive use of dreams and, more largely, of the (human) unconscious. Hence, if the second part of *Agrilogistics* enacts a dream, we might wonder who is then the dreamer: could it be a nonhuman animal dream? Konrad Lorenz, the Austrian father of ethology and a specialist in geese behavior, is supposed to have said that "Geese dream of grain"—meaning *only* of grain—, whereas human dreams, according to Freud's theory, are complex manifestations of the unconscious. In fact, the unconscious is commonly thought of as an exclusively human feature, as Lacan (1974, 15) clearly stated, even though it is associated to the drives, which stand for the most animal part in humans. Philip K. Dick (1968)—one of Apichatpong Weerasethakul's overt references—wondered if androids dream of electric sheep; the second part of *Tropical Malady* could therefore be a dream of the human soldier, or else of the tiger.

One might also wonder what electric sheep dream of. If the second part of *Agrilogistics* is indeed a dream or the unconscious version of the first, the subject of the dream may be the nonhuman animals who appear in this part. There is, though, another possibility: that the machines of the greenhouse facility are the ones who have dreamed. Nowadays, some artificial intelligence specialists state that AI organisms are beginning to have consciousness, as Google engineer Blake Lemoine (2022) claimed, speaking of LaMDA, the company's language model for dialogue applications chatbot development system. Lemoine wished to establish not only that LaMDA had consciousness, but also "sentience." The dialogues with LaMDA that he published on the web show that LaMDA is capable not only of "speaking"—which Nim never learned to do—but also of response-ability, that is, of (apparently) engaging in a relationship with the humans with whom it (or should one say "they"?) interacts—what Nim proved to be very capable of, without needing words. We may wonder if, with consciousness, beings acquire an unconscious too. This perspective opens up new possibilities for human–animal–machine entanglements or hybridizations of which *Agrilogistics* is an outstanding dream.

In conclusion, three of the films discussed here (*Tropical Malady*, *Gente pájaro*, and *Agrilogistics*), although very different, may be included in the category of "shamanic cinema," which encompasses an animist perspective of reality, in the sense of acknowledging agency not only to other-than-human animals—which *Project Nim* also does, although with a more limited scope—, but also to supposedly inanimate objects made by human hands, such as industrial robots. However, shamanic cinema goes beyond this perspective, since it builds bridges between the different worlds that conform this reality, intersecting with each other, like a shaman does. Moreover, as Viveiro de Castro emphasizes, shamanism "presupposes a mode of knowing" (2019, 26). If in modern western epistemology, "to know is to objectify," from an animist standpoint, "to know is to personify, to take on the point of view of that which must be known" (26). Paraphrasing Viveiros de Castro, we could therefore conclude that shamanic cinema "aims at something that is a someone—another subject" (26). Beyond human language, this cinema engages in "transspecific dialogues," standing against the human exceptionalism that has led to the Anthropocene, with the disastrous consequences our entangled worlds are experiencing. Moreover, they tentatively show us how to live, not only in the ruins of a damaged planet (Tsing 2014, 90), but also in the ruins of western modern thought.

Notes

1 Following Thai uses, I will refer to Apichatpong Weerasethakul by his given name.
2 In fact, Bataille wrote "the sexual act," not "desire" (1973, 21).

Works Cited

Agamben, Giorgio. 2002. *L'aperto: L'uomo e l'animale*. Turin: Bollati Boringheri.
Bataille, Georges. 1973. *La Part maudite*. In *Œuvres complètes*, vol. VII. Paris: Gallimard.

Bird-David, Nurit et al. 1999. "Animism Revisited: Personhood, Environment and Relational Epistemology." *Current Anthropology* 40 (S1): S67–S91.
Castro, Carolina. 2021. "Descolonizar los museos desde la espiritualidad y sabiduría mapuche. Entrevista a Francisco Huichaqueo." May 28, 2021. https://artishockrevista.com/2021/05/28/mapuche-entrevista-francisco-huichaqueo/.
Cavalieri, Paola, and Peter Singer, eds. 1993. *The Great Ape Project. Equality beyond Humanity*. New York: Saint Martin's.
Cixous, Hélène, and Jacques Derrida. 2005. *Lengua por venir/Langue à venir. Seminario de Barcelona*, edited by Marta Segarra. Barcelona: Icaria.
Deleuze, Gilles. 1996. "Animal." In *Abécédaire de Gilles Deleuze*, interviews with Claire Parnet. Paris: Vidéo Editions Montparnasse.
Deleuze, Gilles, and Félix Guattari. 1980. *Mille plateaux. Capitalisme et schizophrénie II*. Paris: Éditions de Minuit.
Descola, Philippe. 2005. *Par-delà nature et culture*. Paris: Gallimard.
Díaz Silva, Rafael. 2009. "El nolugar de la música indigenista en Chile: hacia una etnomusicología de la soledad." *Aisthesis* 45, July: http://dx.doi.org/10.4067/S0718-71812009000100012.
Dick, Philip K. 1968. *Do Androids Dream of Electric Sheep?* New York: Doubleday.
Fontenay, Élisabeth de. 1998. *Le Silence des bêtes. La philosophie à l'épreuve de l'animalité*. Paris: Fayard.
Haraway, Donna J. 2003. *The Companion Species Manifesto. Dogs, People, and Significant Otherness*. Chicago: Prickly Paradigm.
Harvey, Graham. 2005. *Animism: Respecting the Living World*. New York: Columbia University Press.
Herzfeld, Chris. 2012. *Petite histoire des grands singes*. Paris: Seuil.
Ingawanij, May Adadol. 2013. "Animism and the Performative Realist Cinema of Apichatpong Weerasethakul." In *Screening Nature: Cinema Beyond the Human*, edited by Anat Pick and Guinevere Narraway, 91–109. New York: Berghahn Books.
Lacan, Jacques. 1974. *Télévision*. Paris: Seuil.
Lamarre, Thomas. 2008. "Animation and Animism." In *Animals, Animality, and Literature*, edited by Bruce Boehrer, Molly Hand, and Brian Massumi, 284–300. Cambridge: Cambridge University Press.
Latour, Bruno. 1997. *Nous n'avons jamais été modernes. Essai d'anthropologie symétrique*. Paris: La Découverte.
Lemoine, Blake. 2022. "Is LaMDA Sentient?—an Interview." June 11, 2022. https://cajundiscordian.medium.com/is-lamda-sentient-an-interview-ea64d916d917.
Mandelbaum, Jacques. 2004. "*Tropical Malady*: traque chamanique dans la jungle du désir." *Le Monde*, November 23, 2004. https://www.lemonde.fr/archives/article/2004/11/23/tropical-malady-traque-chamanique-dans-la-jungle-du-desir_388136_1819218.html.
Marrero-Guillamón, Isaac. 2018. "The Politics and Aesthetics of Non-Representation: Re-Imagining Ethnographic Cinema with Apichatpong Weerasethakul." *Antípoda: Revista de Antropología y Arqueología* 33: 13–32.
Marsh, James. 2011. "James Marsh Talks Project Nim." Interview. *Empire Magazine*. https://www.youtube.com/watch?v=wAPO3ZN1WQs.
Nagel, Thomas. 1974. "What is it like to be a bat?" *The Philosophical Review* 83 (4): 435–50.
Newton, Michael. 2011. *Savage Girls and Wild Boys. A History of Feral Children*. London: Faber & Faber.
Ortín Castellví, Gerard. 2022. "The Scopic Food Regime." In *Cinema in the Anthropocene. Program Participant Abstracts and Bios*. Universitat de les Illes Balears.
Quandt, James, ed. 2009. *Apichatpong Weerasethakul*. Vienna: Österreichisches Filmmuseum.

Segarra, Marta. 2022. *Humanimales: Abrir las fronteras de lo humano*. Barcelona: Galaxia Gutenberg.
Teh, David. 2011. "Itinerant Cinema: The Social Surrealism of Apichatpong Weerasethakul." *Third Text* 25, no. 5 (September): 595–609.
Tsing, Anna Lowenhaupt. 2014. "Blasted Landscapes (and the Gentle Arts of Mushroom Picking)." In *The Multispecies Salon*, edited by Eben Kirksey, 87–109. Durham: Duke University Press.
Viveiros de Castro, Eduardo. 2009. *Métaphysiques cannibales*. Paris: Presses universitaires de France.
Viveiros de Castro, Eduardo. 2019 [2004]. "Exchanging Perspectives: The Transformations of Objects into Subjects in Amerindian Ontologies." *Common Knowledge* 25 (1–3): 21–42.
Weerasethakul, Apichatpong. 2016. *Interview. Our Choices*: https://ourchoices.co/video/apichatpong-weerasethakul/.

10 Being (with) Animals

Human–Horse Relations, Gender, and Queer/Trans Embodiment in Barbara Hammer's *A Horse Is Not a Metaphor* and Ann Oren's *Passage*

Kornelia Boczkowska

Despite experimental filmmakers' ongoing interest in examining the complex ways in which humans and animals relate to each other (Burt 2004, 192–194; Lawrence 2015, 72–86; Knowles 2020, 51–58), closer attention should be paid to how animals and human–animal relations in experimental film speak to gender and queer/trans embodiment. To broach this question, I will discuss how the work of contemporary experimental makers, Barbara Hammer's *A Horse Is Not a Metaphor* (United States, 2008) and Ann Oren's *Passage* (Germany, 2020), makes horses active agents driving the narrative, questioning the conventionality of gendered expression. My discussion responds to the growing body of scholarship on how film and media render animals onscreen, commonly referred to as the animal turn in film studies, pioneered by *Electric Animal* (Lippit 2000) and *Animals in Film* (Burt 2004), and more recently followed by the publication of several books and edited volumes (Pick 2011; Lawrence and McMahon 2015; Smaill 2016; Mills 2017; McMahon 2019; Parkinson 2020; Jackson, Stolley and Payne 2020; Beeler and Beeler 2022; Schultz-Figueroa 2023; Molloy, Duncan and Henry 2023). I am bringing these particular works into discussion because they usefully illuminate each other in terms of their representation of the human–horse, understood as the unified, intercorporeal being, which "draws attention to the transformation of the rider by the horse and the horse by the rider" (Thompson 2011, 223), and overlap with narratives and discourses around the human–animal dualism, using an inherently short, non-verbal and non-narrative format. Despite their different formats and narrative foci, *A Horse Is Not a Metaphor* and *Passage* use the symbolism of the horse, traditionally associated with freedom and the power of living, but avoid anthropomorphization so that animals are given more agency than usual and not only become an object or a metaphor, but also a real subject of the film.

Examples of experimental films, which have abundantly featured horses, not just as mere props or companions but as central characters both closely interlinked and independent of humans, include *Berlin Horse* (Malcolm Le Grice, United Kingdom, 1970), *Man On A Horse* (Corwin Fergus, United States, 1980), *Four Keeps: Horse* (Iichiilish Daik, United States, 2014),

DOI: 10.4324/9781003468028-12

Untitled (Horse) (Rajee Samarasinghe, Sri Lanka, 2014), *3 Dreams of Horses* (Mike Hoolboom, Canada, 2018), *Horsey* (Frederic Moffet, Canada, 2018), *Farming* (Bouge Alexandra, France, 2020), or *My Darling* (Amélie Bargetzi, Switzerland, 2021–2022). Echoing the current trends in animal ethics (Thomas 2016), many such films present animals as more autonomous, self-aware creatures that humans can empathize with (Gruen 2015, 67) rather than absent referents (Adams 1990), which increases animal selfhood and challenges their representation in mainstream media. However, although horses are an integral part of the history of avant-garde cinema, experimental films rarely explore the human–horse boundary in the context of gender divisions, particularly the queer–heterosexual, female–male, and masculine–feminine divide, and allow the characters to physically engage with horses and be less constrained by dominant gender discourses and their own physical limitations. Concerned with human–horse interactions, both films analyzed here directly connect animal bodies to the characters' sick, queer, and trans corporealities, constructing an alternative understanding of the complex bodily entanglements between humans and horses. Blurring rigid boundaries between life forms and species through a gender-bending performance of human–animal hybrids, *A Horse Is Not a Metaphor* and *Passage* bring in the animal viewpoint and questions concerning queer and trans ecology, specifically those that blend certain aspects of ecofeminism and gender transitioning. While both films reflect wider bodies of work and practices of avant-garde cinema through a critical investigation of the ways in which humans and horses relate to each other by means of a highly stylized mise-en-scène and editing strategies, *Horse* queers and de-eroticizes the human–horse, drawing parallels between Hammer's and the horse's aging, fragile bodies, and *Passage* turns it into a beguiling, erotic queer performance, and gender fluid spectacle.

The Gendering of Animals in Experimental Film

Human–animal studies is an emergent and burgeoning interdisciplinary field that explores the ways in which animals intersect with human societies. Ever since John Berger (1977) published his seminal essay, "Why Look at Animals?," the animal turn has continuously challenged the anthropocentrism of the humanities. While cinema and media studies have been rather slow to contribute to animal studies, they have embraced a wide array of perspectives when looking at nonhuman animals, including animal rights activism and animal ethics, postcolonial, feminist, and queer theory, and posthumanism. Although the emergence of animal studies largely coincided with the rise of feminist film theory, it was far less influential than feminism and has gained momentum as a critical paradigm only in the last two decades. Based on the concept of othering, which allows for the assignment of different characteristics to different groups, and the assumption that sexism and speciesism parallel each other, feminist and queer theory links animal studies to gender, often through a materialist perspective. Some of the first studies on ecofeminism included anthropologist Sherry Ortner's (1972) "Is Female to Male as Nature to Culture?," which posits that women, subordinated in most societies, are traditionally associated with nature through childbirth, menstruation, or lactation, and Val Plumwood's (1986) "Ecofeminism: An Overview and Discussion of Positions and Arguments," which critically examines the numerous parallels between the (mis)treatment of women and nature. Similarly, in her theory of the sex-species system, ecofeminist Carol Adams (1990, 90) argues that female and animal bodies, both marked by reproductive abilities, are oppressed and sexualized, especially through meat consumption and pornography, and that the human gaze at animals parallels the male gaze at women. As a result, animals and women can be objectified, ridiculed, fragmented, exploited, and ultimately consumed by men, which obscures the boundary between animal and human. Just like the animalization of Black people and the feminization of the East and Middle-Eastern man through the lens of the Orientalist gaze often linked to Otherness, serving as a means of derision, ridicule, degradation, subordination, oppression, demonization, and orientalism (Bennett 2009, 28), women's animality varies across racial bounds, bringing associations with eroticized and sexualized racial primitivism, a primordial experience of maternity, or patriarchal commodities (Townsend 2024, 489). Similarly, scholars working at the intersection of animal studies, queer studies, and queer ecofeminism (see, e.g., Giffney and Hird 2008; Jagose 2019; Goldsmith 2021), which typically examine gender, sexual, nonreproductive, and (vegan) dietary norms in relation to nonhuman life and non-heteronormativity, have established many connections between the oppression of queer and racialized people (among other groups) and the domination of nature and animals.

On the other hand, through a shared experience of suffering with some humans, animals can be subject to metaphorical displacement and are given multiple meanings, which can result in semantic overload. Consequently, animals in narrative films are not always represented as actual animals, but rather through the lens of anthropocentric constructs, such as anthropomorphism, allegory, and metaphor, and are often signified through their relationship to the human that extends beyond their biological function or physical existence and have no power to affect their signification in human culture. Conversely, *Horse* and *Passage* oppose some of these trends, since the traditional understanding of masculinity and femininity is not necessarily articulated and challenged through othering gender, sexuality, and animality, but instead enacted through human–animal relations, the human–horse, and the animal gaze. The latter concept, originally addressed by Berger (1977) and Derrida (2008), can be understood as "a gaze initiated by the animal, meditative in its quietness and stillness and which compels a response on the part of the human, as it contradicts any assumed superiority of the human over the nonhuman animal" (Woodward 2008, 1). Although *Horse* and *Passage* do not put animals in the observing position, they still recognize the power, complexity, and individuality of the horse gaze. The horse's self-awareness and emotional experience come to the fore, as, rather than being reduced to a symbol, it commands intentionality and agency, questioning human superiority.

Indeed, the gendering of animals in experimental film often reflects Deleuze and Guattari's (2007) concept of becoming animal, which endorses both the Darwinian and zoocentric perspective, resisting comparisons to other species, identities, and power structures (see, e.g., Creed 2009). However, rather than simply reaffirming the affinities between humans and animals, *Horse* and *Passage* disrupt a gendered binary opposition between nature and culture, some traditional gender normative and heteronormative tropes, as well as queer and heterosexual binaries through a gender-bending performance of human–horse hybrids. Since the films enact gender through the interaction between humans and horses, they both perpetuate and challenge the common conception that horses and the horse imagery are gendered and have long been an important expression of hegemonic masculinity shaped by gender identity, social status, the politics of power, and the potential for violence. Birke and Brandt (2009, 189, 196) note that gender in human–horse relationships is expressed "in conjunction with the body and character of the horse" and that riders often seek "a oneness with the horse," which acts as their companion. This suggests that, specifically, horsewomen tend to be less constrained by norms of femininity when being with horses, as this typically connotes masculinity and tomboyishness, ultimately reinforcing a binary. Somewhat opposing this trend, in *Horse* and *Passage*, gender constructs are loosened as human characters and horses are connected through embodiment and create a unique form of intimacy and mutual corporeality, one that is enacted through the human and horse's body and helps perform the characters' (including horses') fears, desires, and gender identity in various ways (Birke and Brandt 2009, 195). However, rather than removing the horse from the regimes of anthropomorphism or symbolism, common for the materialist film practice (Lawrence 2015), both works highlight its ambiguous status, yet at the same time grant the animal a thoroughly recognizable form and allude to conventional meaning-making processes, demonstrating an attentiveness to the characters' close emphatic relationship with animals.

Being with Horses in *A Horse Is Not a Metaphor*

Horse is a thirty-minute multilayered autobiographical documentary and essay film, directed by widely celebrated American feminist and queer maker Barbara Hammer and scored by experimental vocal artist Meredith Monk.[1] Usually interpreted as "a touching portrait of a body that has given up its pretenses but not its desire to live" (Osterweil 2010, 198), which expresses "the desire for cancer survivorship" (Stacey and Bryson 2012, 10), the film stands out from the artist's oeuvre, which often privileges sexual content and lesbian themes through its focus on some fundamental questions related to human–animal relations, aging, and illness. *Horse* won the 2nd Prize at the Black Maria Film Festival in 2009 and documents Hammer's fight against ovarian cancer through combining footage of her chemotherapy sessions with images of her horseback riding, accompanied by dialogue between the artist and healthcare providers, Meredith Monk's haunting, poignant musical composition as well as occasional voiceover narration and intertitles. Diagnosed with stage three ovarian cancer, Hammer had been for over two years in remission from the disease at the time of shooting, but she described herself as a thriver rather than survivor due to a seventy percent chance of recurrence within two years. In the film's synopsis, Hammer (2008) states that "lesbians increase their risk for this disease if they have not had children and have infrequent gynecological exams due to the continuance of a homophobic medical profession [and] it is also a difficult disease to detect before metastises," which links the oppression of queer people to the oppression of animals since the horses the artist bonds with have been also diagnosed with cancer and are subject to the rider's will.

Despite this confession and a difficult medical diagnosis, *Horse* takes a rather optimistic tone conveying a vision of a lesbian living with cancer who managed to maintain a "positive attitude of hope," celebrate her "exaltation and love of life" and ultimately flourish, particularly when being with horses (Hammer 2008). While Hammer's partner, Florrie, was taking stills and videotaping the artist's progression when she was undergoing extensive chemotherapy, Hammer's friend and filmmaker from San Francisco, who took her for a trip to the country for a week of retreat, shot her riding horseback, swimming in a Catskill creek, and walking naked through the woods. These images were later superimposed over footage Hammer shot herself from her hospital bed, which depicted the filmmaker's steroid-swollen face, hypodermic needles, chemotherapy bags hanging above her bed, and the nurses attending her. In the film's synopsis, Hammer (2008) admitted that she used horse images as a source of inspiration and meditation when she was undergoing therapy at the hospital as they took her "out of the confines of the hospital room and to a landscape that knew no boundaries." *Horse* captures the moments of recovery and normality, which followed the remission of the disease, using footage of some intimate aspects of the medical procedures Hammer undertook at the hospital, footage of horse riding at a Wyoming ranch, and footage of the two horses, Casper and Josephine, who—quite significantly—were simultaneously ill with tumors at the time of shooting.

The trip to the country partly fulfilled Hammer's childhood and lifelong dream of having a horse in her backyard, and reviving it every day through looking at horse images at the hospital literally kept the artist alive for four and a half months of rigorous chemotherapy sessions. Although the dream did not come true per se, Hammer was fortunate enough to have some friends and strangers generously let her ride their horses and explore the trails of "the red hills of Georgia O'Keefe's Ghost Ranch in New Mexico, the grassy foothills of the Big Horn in Wyoming, and leafy paths in Woodstock, New York" (Hammer 2008). While the film's structure evokes Maya Deren's vertical filmmaking, through the use of superimpositions and layering of images

(Hammer 2010), which shows the transformative power of death and rebirth, the camera acts as a means of recovery from the torments of chemotherapy and the horse literally and figuratively changes illness into recovery. Although the film contains scarce dialogue, incorporated in lyrical phrases describing the artist's inner experiences, and interspersed with textbook descriptions of the symptoms tormenting her, selected phrases and utterances indicate the enactment of the human–animal hybrid, such as "I want to be a bird migrating" or "Horse and I are one." As Hammer is often depicted bathing with her dog Spooner or filming from horseback, her close relationship with animals is primarily signaled on the level of the image, particularly regarding her interactions with horses. According to the filmmaker, the heroine and the horse are connected through embodiment, the senses, the freedom of movement, and the power of living to the fullest in their mutual quest for life: "Freedom is my eye and mind following the flow of expression through movement. Freedom is riding my horse on a trail exploring the unknown and seeing with fresh eyes as the world becomes new again." Throughout the film, the artist continuously juxtaposes point-of-view shots of the two horses' bodies, stricken by illness, her own body, and her petting the horse or standing right next to it, and the camera occasionally closes up on the horse's eyes and profile, sometimes superimposed over Hammer's eyes and face, which illuminates the animal gaze and the filmmaker's naked, bruised, winkled, swollen, and balding body, presented as both diseased and desiring (Stacey and Bryson 2012, 8).

While slow-motion shots of women rodeo-riders managing their stallions, juxtaposed with footage of Hammer's reluctant submission to the next round of chemotherapy treatment, allude to some traditionally masculine skills, the oppression of animals, and the artist's lesbian identity, I suggest that the human–horse in *Horse* stands for queer embodiment and both the fragility and strength of the artist's and horse's body, driven by illness and a desire to live. Similarly, a close-up shot of Hammer's bald head and regrowing white hair, superimposed over the image of a galloping horse and the grey coat of another horse, respectively, or footage of the artist's and horse's legs standing next to each other, further transcend the human–animal boundary, evoking a highly tactile, haptic sense of co-presence through highlighting the textures of the human and animal bodies. Although the title playfully implies that the horse is not just a metaphor, but, as put by Hammer (2008), "a living, breathing creature of power and pride that [she] join[s] within moment-by-moment living," the animal still emerges as a metaphor for "the emotional ups and downs of a cancer patient." At the same time, the horse experiences disease and undergoes treatment just like the artist herself, which further encourages boundary-crossing and the harmonization of human and animal bodies. In their essay on queering cancer survivorship, Stacey and Bryson (2012, 8) suggest that *Horse* speaks directly to the statement Sontag (1978) made in her widely acclaimed book, *Illness as Metaphor*, about the power of the metaphorical designation in the stigmatization of sexuality and illness.

Consequently, they argue, while the horse is not a metaphor, but stands for the materiality of embodiment determined by life, illness, and sexuality, it still seems impossible to strip illness and animals of their symbolic force (Stacey and Bryson 2012, 9). For instance, the synergy of human and animal is symbolically catalyzed by images of water and nature, which have a purifying, healing, and cleansing power, bringing about rebirth and renewal. This is captured in one of *Horse*'s most well-known scenes in which Hammer is curling in a fetal manner toward a vortex of water and light, followed by the image of a rearing horse that conveys a strong sense of terror and rebellion against the cancer diagnosis.

Hammer's repeated use of montage, superimpositions, editing rhythms, slow motion, juxtapositions, repetition, image re-scanning, optical printing, extreme close-ups, re-assemblage, and the first-person voice-over narration not only evoke a sense of beauty, awe, and wonder that the artist experienced when she was shooting the varied, vibrant natural landscape from horseback, but corresponds directly to her multilayered experiences, emotions, and feelings, ranging from hopes and thrills to fears and despair. However, rather than perpetuating the myth that riding horses grants women freedoms that they may be denied, helps boost their self-esteem and confidence, and challenges some heterosexual assumptions and conventional concepts of femininity (Traeen and Wang 2006), Hammer draws the analogy between horse riding and being a lesbian fighting cancer by exposing her skinny body, bald head, and her sexuality. Instead of showcasing Barbara's high levels of skill required to practice horsemanship, which are traditionally associated with masculinity, *Horse* articulates lesbian identity through human–animal relations and queer ecofeminism. The fact that the heroine engages in nature-oriented behaviors (swimming naked in a river, bathing in natural surroundings, playing with her dog Spooner, riding horseback) and a non-abusive relationship with horses echoes one of the central premises of queer ecofeminism, which envisions a form of ecotopia where women, animals, queer people, and nature are both liberated and emancipated, co-existing peacefully and harmonically (Ourkiya 2023, 120). Subtly linking to queer ecofeminism, the film acknowledges connections between the oppression of sexuality, indirectly hinted at in Hammer's diagnosis, and the oppression of nature, evident in the frailty of the horse's and Hammer's body, their struggle with cancer and Hammer's attempt to ride (or sometimes tame) horses as a means of recovery from the torments of chemotherapy. *Horse* disrupts the natural-unnatural dualism and emphasizes queer people's connection toward environment through sequences featuring riding horses and Hammer's naked body surrounded by or immersed in nature, including lakes, mountains, and woods, which also exemplifies the artist's lifelong practice of representing lesbian sexuality on the screen (Stacey and Bryson 2012, 9). In doing so, it "reconfigures the dynamic vitalities of human and horse" (Stacey and Bryson 2012, 9), drawing attention to the human–horse, which creates new connections across the human and animal world as it places a stronger emphasis on a woman's and horse's aging, sick bodies, and interconnectedness, expressed by the use of juxtapositions, superimpositions, and multilayered images.

Becoming Horse in *Passage*

Compared to Hammer's film, Israeli-born, Berlin-based artist Ann Oren's 16 mm *Passage* still places queer embodiment at the center of the narrative, but is more concerned with the human trans body's entanglement with the nonhuman and the former's metamorphosis, which stems from the interaction between the two characters.[2] In the film, the protagonist's queer and trans corporeality "indicates movement across different sites" (Alaimo 2017, 238), expressed through their imitation of the horse's physical movements and their attempt to establish a dialogue and a close, almost intimate relationship with the horse. Echoing the key assumptions of trans ecology, which examines questions of nature, gender, and trans embodiment (Seymour 2020, 191), the character engages in the trans movement, turning into and merging with the animal through a smooth, gradual, and tactile process of physical bodily communication and interaction between human and nonhuman. While the connections between LGBTQ+ issues, queer theory, and ecocriticism have been theorized by Gaard (1997), Mortimer-Sandilands (2005), and Huang (2020), Seymour (2013, 36) has developed a counter discourse termed "organic transgenderism," which *Passage* seems to seek inspiration from in its construction of gender fluidity. The concept denotes "a vision of gender transitioning as spontaneous, immanent, self-driven, and generally 'natural'—akin to the life-cycle changes of plants and animals—in opposition to its dominant framing in popular and medical discourse as an 'unnatural,' technoscientific intervention" (Seymour 2020, 191). Building on ecocriticism, ecofeminism, the new materialisms, material ecocriticism, transgender studies, queer ecology, and environmental and trans justice studies, trans ecology is primarily concerned with exploring non-normative gender, contesting dualisms and binaries, especially regarding the heterosexual/queer and culture/nature dichotomies, and drawing parallels between the risks posed to the environment and trans persons. When linking to animal studies and the Anthropocene, trans ecology further blurs the line between human and nonhuman, laying the foundations for trans-animal studies (Stryker and Aizura 2013; Sundén 2015) and tranimalities (Hayward and Weinstein 2015; Steinbock, Szczygielska and Wagner 2021) and raising questions about species extinction and trans embodiment (Woelfle-Erskine and Cole 2015).

In view of trans ecology, gender and nature are inextricably bound together and humans are equal to animals once they give up privilege and cease to elevate themselves over others. However, Bedford (2020, 1) also notes that since nature has been traditionally associated with the lower-ranked identifiers of binaries, including culture/nature, man/woman, or mind/body, such associations have often been detrimental to women, racialized people, differently gendered individuals, the disabled, the poor, and other groups discriminated against on the basis of their bodies and assumed proximity to nature. Trans ecology challenges the allocation of nature and animals as other and deconstructs hierarchies that view transgender identity as unnatural, particularly for those with trans identities and attempting to reclaim nature, seen as inherently heteromasculine, "from the grasp of masculinity and heteronormativity" (Thorsteinson and Joo 2020, 33), which is sometimes considered radical as transgender people have been historically constructed as being unnatural (Bedford 2020, 1). At the intersection of queer and environmental studies, trans is usually defined as "the movement across a socially imposed boundary away from an unchosen starting place—rather than any particular destination or mode of transition" (Stryker 2008, 1), which implies that the trans movement or transition is a natural rather than unnatural process (Stryker 2020, xviii). However, as I will show, what seems to be particularly relevant for *Passage* is the concept of trans-corporeality, originally coined by Alaimo (2010) and based on the assumption that all humans are trans and intermeshed with both their own environments and the nonhuman world. While this is not without controversy as it elides the specific transgender experience and suggests a more utopic trans experience, one that does not entail the violence many transgender individuals face, trans-corporeality in Alaimo's (2017, 435) view and in *Passage* simply means that all living, embodied creatures are intermeshed with the material world, which disrupts Western human exceptionalism and individualism, highlighting the permeability of the human–animal boundary. *Passage* additionally resonates with a prepositionally oriented notion of trans, understood in its prefixial state and ontological terms as a process, movement, and expressive force through which beingness is constituted (Hayward and Weinstein 2015). This echoes one of the chief tenets of tranimalities, which emerges at the intersection of trans theory and animal theory (Derrida 2008; Grosz 2010; Chen 2012; Steinbock, Szczygielska and Wagner 2021) and views transing as "the practice that takes place within, as well as across or between, gendered spaces" (Stryker, Currah, and Moore 2008, 13). In *Passage*, the human–horse functions as a form of trans ecological kinship that invites affect, intimacy, and care, and in which the trans body of the human character enters into a queer interspecies relationship with the animal through a process of imitation and appropriation.

Passage features gender fluid foley artist Simon(e) Jaikiriuma Paetau, starring as a dressage horse, who creates sounds for the film. As Oren (2021) states, *Passage* alludes to Eadweard Muybridge's pre-cinematic experiments with horses and focuses on the process of transformation of the performer into a gender-defying centaur, often associated not only with virility, masculinity, and male sexual drive, but also with femininity, sensitivity, and dance choreography (Segarra 2020, 872). Originally conceived and exhibited as a video installation and scored by Manuela Schininá, the film is a relatively loose montage of several scenes in which a naked Simon(e) mimics the animal's tail-swish, hoof clip clops, and snorts with their body, props, and ponytail attached to their tailbone, for example, through knee walking, licking their lips, pulling the chain around their neck, shaking their head and fake ponytail, knocking their knees, or hitting the sand with coconut shells split into two parts. Each scene in which Simon(e) imitates the horse is immediately followed by a scene in which the actual horse, ridden by a woman jockey in a horse-riding arena, repeats the same gestures, movements, and sounds, that is, tail-swishes, clip clops, and snorts. In her attempt to create the portrait of the horse's and artist's body, Oren deliberately juxtaposes Simon(e)'s performance with the horse's real-life performance shot by shot, which augments the film's associative potential as well as its olfactory, haptic, and tactile quality. Since Simon(e) defines themself as queer and non-binary, *Passage* reflects not only on the human–animal boundary, as it creates an intimacy between the two characters through the embodiment of the horse's movements and sounds, but also on fictional gender roles and gender fluidity. Oren (2021) sees Simon(e) as a neo-femme fatale figure, a self-determining, pro-active person who subverts patriarchal principles and gender norms (Piotrowska 2018), and encourages her viewers to think about the role of horses and gender in cinema history, going back to Muybridge's *The Horse in Motion*, seen as one of the first instances of moving images.

In *Passage*, Oren is primarily concerned with the process of imitation, as the foley artist mimics the horse to the point where they become viscerally connected and the performer themself melts into their own work. Simon(e)'s unique on-screen presence, charm, elegant burlesque glamor, and distinct bodily movements are almost erotic and have a very strong magnetism, which establishes the supremacy of flesh over spirit and empowers the act of gazing at the human–animal body by depicting it in motion. Also, nudity in *Passage* continually disturbs the viewer's habitual gaze and transgresses gender norms as the body represents the human–animal's corporeal functioning, acting both as an erotic object of the gaze and a transformative bodily practice. This changes in the closing scene when Simon(e) puts on a white coat, which partly covers the ponytail, and slowly walks into the theatre, climbs the stairs, and sits in one of the middle rows in an empty, dimly lit auditorium, staring into the screen. When captured from behind, Simon(e) can be seen petting their ponytail and watching the human hand combing the horse's ponytail on the screen, which is followed by two extreme close-up shots of Simon(e)'s and the horse's eyes that never meet. The scene connotes less eroticism and seduction and instead shifts our attention to the gaze, which has been featured in two previous scenes where the camera closes up on the horse's legs and Simon(e)'s hand petting the horse's body, both reflected in the foley artist's eye. Although Simon(e) is unable to overcome the boundaries of the human world, their attempt to mimic the movements of the horse and interactions with the animal bridges the human–nonhuman divide and allows the characters to share a common ground that leaves the viewer uncertain if they just play horse or really engage in the process of becoming(with) horse. *Passage* stresses the fluidity of transition between human and nonhuman and consequently creates a mesmerizing, yet also somewhat disturbing human–animal hybrid established through transpecies intimacy and a physical proximity between Simon(e) and the horse, which further points to naturalness and authenticity of trans embodiment. Blurring the boundaries between human and animal, *Passage* illustrates the concept of organic transgenderism as it creates strong haptic and aural sensations of becoming horse through sounds and images, challenging fictive gender roles in film history as the viewer observes Simon(e) play with traditional gender expectations and transform from human to animal.

Conclusion

Both *A Horse Is Not a Metaphor* and *Passage* bring attention to human–horse mobility and bodily movements as well as human interactions with horses, particularly Hammer's horse riding and Simon(e)'s impersonation of the horse's behavioral patterns. This directly links the animal to the characters' performance of femininity, queerness, and gender fluidity, giving prominence to the human–horse, understood as a symbiotic organism and an intercorporeal being of human and animal connected through embodiment, which functions as an extension of the horse's body into the human and vice versa. However, rather than simply demonstrating the rider's appropriation of the horse's body as an extension of their own, the films question the horse as a metaphor and symbol of stereotypical masculinity and—ultimately—the binary opposition between masculinity and femininity, freeing the human characters from constrictive gender norms (Segarra 2020, 883). While often presenting all characters in movement, they also expose the human and animal gazes and bodies in near proximity to each other, highlighting the close relationship between humans and animals. Being with, riding, and moving like horses enables a physical and spiritual transformation and allows for gender performance, specifically for the enactment of femininity and queerness, which is deeply corporeal and plays out in many forms, becoming a meaningful bodily practice. While *Horse* pays homage to Hammer's ongoing battle with ovarian cancer, her process of recovery, and the fragility of the human–horse, *Passage* speaks to gender fluidity and the boundaries between humans and animals.

In *Horse*, Hammer performs being with horses and with cancer in a way that is closely attuned to queer practices of freedom and embodied trauma, challenging women's relationship with horses, often informed by women's misplaced or unfulfilled desires, particularly a desire for a man and for a child (Birke and Brandt 2009, 190), here replaced simply with the artist's desire for living. However, rather than solely narrating and choreographing Hammer's cancer experience, *Horse* deploys a range of experimental editing techniques to poetically document the artist's chemotherapy treatments, walking, swimming, and horse riding, juxtaposed with images of medical technologies, horses, and parts of the human body, particularly extreme close-ups of the characters' skin, hair, and coat. This prioritizes the human–horse, human–animal mobility, and the act of riding in which human and horse are closely interlinked and become an embodiment of each other, capable of transcending or subverting social boundaries, mostly related to aging and illness, and traditional gendered divisions and modes of conduct. While the clinic segment of the film is intensely private and personal, due to the frequent use of the hand-held camera and close-up shots of medical procedures and the intraperitoneal chemotherapy Hammer is subjected to, encouraging the viewer to voyeuristically gaze at her treatment and body, the being with horses and horse-riding sequences draw attention to animals in freedom, captivity, well-being, and distress. The images of wild, unrestrained horses in wide-open spaces are often juxtaposed with footage depicting the effects of treatments on what the filmmaker describes as her poisoned body, including weight and hair loss and surgery scars (Juhasz 2019, 94), which not only strips her body of both a queer erotic appeal and femininity, but also highlights the co-presence of human and animal, evident in horse riding and images of the human and animal characters' superimposed eyes, hair, profiles, posture, gestures, and stance.

Similarly, *Passage* foregrounds human–horse vitalities, but at the same time shifts the focus from the physicality of illness, a diseased, damaged female body, and a lesbian desire to live to a graceful beauty and bodily symbiosis of the human-horse, transpecies intimacy, and queer embodiment. Simon(e)'s body-based performance engages with questions of animal–human interactions, the animal gaze, and material, kinesthetic affinities between human and horse in the context of transgender identity and transformation, somewhat fulfilling what Hilton (2013, 491) calls the artists' longstanding fantasy of hybridization between human and horse. However, rather than augmenting the human body through incorporating the animal other, *Passage* presents Simon(e) engaging in a theatrical mimetic representation of the horse's bodily movements and sounds and absorbing the attributes of the nonhuman, which attests to the fragility and fluidity of inter-species barriers, especially when human and animal come together in movement. Through its sensitive exploration of the human hybridity with other animals, Oren's film queers the human–animal boundary, enhances a subtle eroticism of the human–horse, and expands our understanding of transgender embodiment that moves beyond the human subject, playing with gender fluidity and positioning the human–horse as the boundary-defining other. Simon(e)'s embodied performance emerges as a gradual, haptic process of coming into contact with horses and the ways in which they are similar to and distinct from humans, which becomes evident in the latter part of the film when the foley artist watches a mere on-screen representation of their interactions with the horse. Interestingly, Simon(e)'s act also reverses the traditional role of performing animals through centering on the human improvisation of the horse's behavior (see, e.g., Tait 2015), implying that the process of (trans)formation of human into nonhuman also involves the horse's side of experience of being with humans, often overlooked in modern art and narrative film (Hilton 2013, 488; Cull 2015, 33). While *Passage* posits that becoming animal should be approached creatively through performative practices, it also reminds us—especially in the final scene when Simon(e) leaves the theatre and the lights go off—that the human will never be able to fully identify and sympathize with the animal and comprehend its thinking patterns and way of living and any attempts to control it or learn its language will most likely fail.

It seems that *Horse* and *Passage* not only link queer and trans embodiment to the movement of the human body, which seems to fuse with that of the animal, giving rise to the human–horse, but are also concerned with the horse's encounter with humans. In the films' finale, Hammer draws her face extremely close to the horse's head, looking into its eyes, and the horse looks back at the artist, and Simon(e), acting as a dressage horse, eventually dissolves into their own imitation. It becomes clear, then, that while *Horse* and *Passage* attempt to bridge the human–animal divide, horses do not function as mere props or companions, but only merge with humans and affect their experience of life, disease, and gender. In doing so, the films subvert human dominance over the natural world as they expose the characters' weakness and fragility in confrontation with illness, restrictive gender expectations, and horses themselves, empowering animal agency. As Simon(e)'s seductive, yet still imperfect imitation of the horse and Hammer's mastering of Casper and Josephine may appear to reinforce human speciesism, *Horse* and *Passage* also make a strong statement that horses are not subordinated to human power. Instead, as the characters identify with animals not only on a physical, but also emotional level in their struggle with ovarian cancer, aging, trans identity, and normative gender norms, horses are shown as capable of shaping interspecies engagements and actions and function as independent beings rather than just metaphors or companions for humans.

Funding

This work was funded by the National Science Centre, Poland, under the project: "An animal is not a metaphor: Animals and human–animal relations in American experimental ecocinema" (grant no. 2023/49/B/HS2/00653).

Notes

1 *A Horse Is Not a Metaphor* is available for viewing at: https://ubu.com/film/hammer_horse.html. Barbara Hammer (1939–2019) was an award-winning and widely celebrated feminist experimental filmmaker, writer, and pioneer of queer and lesbian cinema whose career spanned over fifty years. Since the late 1960s, she made over 90 moving image works, including 16 mm films and videos, performances, photographs, drawings, installations, and collages, which reflect on gender roles, lesbian and queer relationships, family, aging, and death. Her work is sometimes considered controversial due to its explicit representation of taboo issues, such as sexuality, orgasm, menstruation, and lesbianism (Osterweil 2010, 185; Juhasz 2019, 102).

2 *Passage* is available for renting at https://vimeo.com/ondemand/passage3 and the trailer can be viewed on Ann Oren's official website: https://www.annoren.com/copy-of-blue. Ann Oren (born 1979) is an Israeli artist and award-winning filmmaker based in Berlin. Since 2010, she has completed several artworks and experimental films, which explore the relationship between humans, animals, plants, and the broadly defined ecosystem, particularly in the context of gender identity, animality, fictosexuality, interspecies relations, and digital culture.

Works Cited

Adams, Carol J. 1990. *The Sexual Politics of Meat: A Feminist-Vegetarian Critical Theory*. New York: Continuum.

Alaimo, Stacy. 2010. *Bodily Natures: Science, Environment, and the Material Self*. Bloomington: University of Indiana Press.

Alaimo, Tracy. 2017. "Trans-Corporeal Feminisms and the Ethical Space of Nature." *Revista Estudos Feministas* 25 (2): 909–934.

Bedford, Anna. 2020. "Introduction. Transecology—(Re)Claiming the Natural, Belonging, Intimacy, and Impurity." *Transecology: Transgender Perspectives on Environment and Nature*, edited by Douglas A. Vakoch, 1–16. New York: Routledge.

Beeler, Karin, and Stain Beeler. 2022. *Animals in Narrative Film and Television: Strange and Familiar Creatures*. Lanham: Rowman & Littlefield.

Bennett, Joshua. 2009. "Buck Theory." *The Black Scholar: Journal of Black Studies and Research* 49 (2): 27–37.

Berger, John. 1977. "Why Look at Animals?" In *About Looking*, 3–28. New York: Pantheon Books.

Birke, Lynda, and Keri Brandt. 2009. "Mutual Corporeality: Gender and Human/Horse Relationships." *Women's Studies International Forum* 32 (3): 189–197.

Burt, Jonathan. 2004. *Animals in Film*. London: Reaktion Books.

Chen, Mel Y. 2012. *Animacies: Biopolitics, Racial Mattering, and Queer Affects*. Durham: Duke University Press.

Creed, Barbara. 2009. *Darwin's Screens: Evolutionary Aesthetics, Time and Sexual Display in the Cinema*. Melbourne: Melbourne University Press.

Cull, Laura. 2015. "From Homo Performans to Interspecies Collaboration: Expanding the Concept of Performance to Include Animals." In *Performing Animality Animals in Performance Practices*, edited by Lourdes Orozco and Jennifer Parker-Starbuck, 19–38. Basingstoke: Palgrave Macmillan.

Deleuze, Gilles, and Félix Guattari. 2007. "Becoming Animal." In *The Animals Reader: The Essential Classical and Contemporary Writings*, edited by Linda Kalof and Amy Fitzgerald, 37–50. Oxford: Berg.

Derrida, Jacques. 2008. *The Animal that Therefore I Am*. Translated by David Wills. New York: Fordham University Press.

Gaard, Greta. 1997. "Toward a Queer Ecofeminism." *Hypatia* 12 (7): 114–137.

Giffney, Noreen, and Myra J. Hird, eds. 2008. *Queering the Non/human*. Aldershot: Ashgate.

Goldsmith, Mitch. 2021. "Queering Our Relation with Animals: Multispecies Sexuality beyond the Laboratory." *Gender and Sexuality in Critical Animal Studies*, edited by Amber E. George, 45–62. Lanham: Lexington Books.

Grosz, Elizabeth. 2010. *Becoming Undone: Darwinian Reflections on Life, Politics, and Art*. Durham: Duke University Press.

Gruen, Lori. 2015. *Entangled Empathy*. New York: Lantern Books.

Hammer, Barbara. 2008. Synopsis of *A Horse Is Not a Metaphor*. https://barbarahammer.com/films/a-horse-is-not-a-metaphor/.

Hammer, Barbara. 2010. "Modern Women: Barbara Hammer on Maya Deren." *The Museum of Modern Art*. https://www.youtube.com/watch?v=3pTVbQilDqY.
Hayward, Eva, and Jami Weinstein. 2015. "Introduction: Tranimalities in the Age of Trans* Life." *TSQ: Transgender Studies Quarterly* 2 (2): 195–208.
Hilton, Leon J. 2013. "The Horse in My Flesh: Transpecies Performance and Affective Athleticism." *GLQ* 19 (4): 487–514.
Huang, Peter I-min. 2020. "Chinese Literature, Ecofeminism, and Transgender Studies." In *Transecology: Transgender Perspectives on Environment and Nature*, edited by Douglas A. Vakoch, 97–109. New York: Routledge.
Jackson, Kathy, Kathy Stolley, and Lisa Payne, editors. 2020. *Animals and Ourselves: Essays on Connections and Blurred Boundaries*. Jefferson: McFarland.
Jagose, Annamarie. 2019. "Anthropomorphism, Normativity, and the Couple: A Queer Studies/Human-Animal Studies Mash-Up." *GLQ: A Journal of Lesbian and Gay Studies* 25 (2): 315–335.
Juhasz, Alexandra. 2019. "Coda: In Love, Anger, and Loss: Barbara Hammer and Carolee Schneemann." *Feminist Media Histories* 5 (4): 87–105.
Knowles, Kim. 2020. *Experimental Film and Photochemical Practices*. Cham: Palgrave Macmillan.
Lawrence, Michael. 2015. "Muybridgean Motion/Materialist Film: Malcolm Le Grice's Berlin Horse." In *Animal Life and the Moving Image*, edited by Michael Lawrence and Laura McMahon, 72–94. London: Palgrave Macmillan.
Lawrence, Michael, and Laura McMahon, eds. 2015. *Animal Life and the Moving Image*. London: Bloomsbury.
Lippit, Mizuta. 2000. *Electric Animal: Toward a Rhetoric of Wildlife*. Minneapolis: University of Minnesota Press.
McMahon, Laura. 2019. *Animal Worlds: Film, Philosophy and Time*. Edinburgh: Edinburgh University Press.
Mills, Brett. 2017. *Animals on Television: The Cultural Making of the Non-Human*. London: Palgrave Macmillan.
Molloy, Missy, Pansy Duncan, and Claire Henry. 2023. *Screening the Posthuman*. Oxford: Oxford University Press.
Mortimer-Sandilands, Catriona. 2005. "Unnatural Passions? Toward a Queer Ecology." *Invisible Culture* 9, www.rochester.edu/in_visible_culture/Issue_9/title9.html.
Oren, Ann. 2021. "Intro: Ann Oren: Passage im KINDL." *KINDL – Zentrum für zeitgenössische Kunst*. https://www.youtube.com/watch?v=ZOGWpPr2niY.
Ortner, Sherry B. 1972. "Is Female to Male as Nature Is to Culture?" *Feminist Studies* 1 (2): 5–31.
Osterweil, Ara. 2010. "A Body Is Not a Metaphor: Barbara Hammer's X-Ray Vision." *Journal of Lesbian Studies* 14(2–3): 185–200.
Parkinson, Claire. 2020. *Animals, Anthropomorphism and Mediated Encounters*. London: Routledge.
Pick, Anat. 2011. *Creaturely Poetics: Animality and Vulnerability in Literature and Film*. New York: Columbia University Press.
Piotrowska, Agnieszka. 2018. *The Nasty Woman and The Neo Femme Fatale in Contemporary Cinema*. London: Routledge.
Plumwood, Val. 1986. "Ecofeminism: An Overview and Discussion of Positions and Arguments." *Australasian Journal of Philosophy* 64: 120–138.
Ourkiya, Asmae. 2023. *Queer Ecofeminism: From Binary Environmental Endeavours to Postgender Pursuits*. London: Rowman & Littlefield.
Schultz-Figueroa, Benjamin. 2023. *The Celluloid Specimen: Moving Image Research into Animal Life*. Berkeley: University of California Press.
Segarra, Marta. 2020. "Performing Centaurs and the Debasement of Masculinity." *Men and Masculinities* 23 (5): 872–888.

Seymour, Nicole. 2013. *Strange Natures: Futurity, Empathy, and the Queer Ecological Imagination*. Champaign: University of Illinois Press.
Seymour, Nicole. 2020. "Good Animals: The Past, Present, and Futures of Trans Ecology." In *Transecology: Transgender Perspectives on Environment and Nature*, edited by Douglas A. Vakoch, 190–204. New York: Routledge.
Smaill, Belinda. 2016. *Regarding Life: Animals and the Documentary Moving Image*. Albany: SUNY.
Sontag, Susan. 1978. *Illness as Metaphor*. New York: Farrar, Straus & Giroux.
Stacey, Jackie, and Mary Bryson. 2012. "Queering the Temporality of Cancer Survivorship." *Aporia* 4 (1): 5–17.
Steinbock, Eliza, Marianna Szczygielska, and Anthony Clair Wagner. 2021. *Tranimacies: Intimate Links Between Animal and Trans* Studies*. New York: Routledge.
Stryker, Susan. 2008. *Transgender History*. Berkeley: Seal.
Stryker, Susan. 2020. "Foreword." In *Transecology: Transgender Perspectives on Environment and Nature*, edited by Douglas A. Vakoch, xvi–xix. London: Routledge.
Stryker, Susan, and Aren Z. Aizura, 2013. "Introduction: Transgender Studies 2.0." In *The Transgender Studies Reader 2*, edited by Susan Stryker and Aren Z. Aizura, 1–12. New York: Routledge.
Stryker, Susan, Paisley Currah, and Lisa J. Moore. 2008. "Introduction: Trans-, Trans, or Transgender?" *WSQ: Women's Studies Quarterly* 36 (3–4): 11–22.
Sundén, Jenny. 2015. "On Trans-, Glitch, and Gender as Machinery of Failure." *First Monday* 20 (4): https://firstmonday.org/ojs/index.php/fm/article/view/5895.
Tait, Peta. 2015. "Acrobatic Circus Horses: Military Training to Natural Wildness." In *Performing Animality Animals in Performance Practices*, edited by Lourdes Orozco and Jennifer Parker-Starbuck, 97–116. Basingstoke: Palgrave Macmillan.
Thomas, Natalie. 2016. *Animal Ethics and the Autonomous Animal Self*. Cham: Springer.
Thompson, Kirrilly. 2011. "Theorising Rider-Horse Relations: An Ethnographic Illustration of the Centaur Metaphor in the Spanish Bullfight." In *Theorising Animals*, edited by Nicola Taylor and Tania Signal, 221–253. Leiden: Brill.
Thorsteinson, Katherine, and Hee-Jung Serenity Joo. 2020. "Coming Out, Camping Out. Transparent's Eco-ethical Approach to Gender." In *Transecology: Transgender Perspectives on Environment and Nature*, edited by Douglas A. Vakoch, 31–53. London: Routledge.
Townsend, Karim. 2024. "On Mike White's Primitivist Posthumanisms: Animality, Coloniality, and Racial Affect in *The White Lotus*." *Quarterly Review of Film and Video* 41 (4): 487–514.
Traeen, Bente and Catherina E. Wang. 2006. "Perceived Gender Attribution, Self-esteem and General Self-efficacy in Female Horseback Riders." *Journal of Equine Veterinary Science* 25 (10), 439–444.
Woelfle-Erskine, Cleo, and July Cole. 2015. "Transfiguring the Anthropocene: Stochastic Reimaginings of Human-Beaver Worlds." *TSQ: Transgender Studies Quarterly* 2 (2): 297–316.
Woodward, Wendy. 2008. *The Animal Gaze: Animal Subjectivities in Southern African Narratives*. Johannesburg: Wits University Press.

11 Biological Imagination, Critical Environmentalism, and Anthropocene in *Annihilation*

Galyna Maleeva and Camil Ungureanu

Introduction

Critically thinking about human–nature relations in the Anthropocene generates the urgent need for the decentering and transformation of conventional humanism. In one perspective, "the Anthropocene has become a highly mediated assemblage of discourses and phenomena regulating relations between humans and the earth as a central focus point for subjectivities in transformation" (Bee 2020, 191; see also Tønder 2017; Bonneuil and Fressoz 2016; Clot Garell 2024). More ambitiously, the onset of the Anthropocene requires a theoretical and practical shift toward reimagining subjectivity and nature, redefining communal connections and non-anthropocentric kinships, and advocating for transspecies justice. These transformative imperatives bear extensive implications, both in terms of ontology and socio-political dynamics. From a decentering perspective advanced in social sciences, arts, and critical environmental studies, rational subjectivity and societal relations are rethought as deeply embedded within relations of affective interdependence with and care for the environment (Wapner 2008; Pellizzoni, Leonardi, and Asara 2022; González-Hidalgo and Zografos 2020; Beistegui 2025; Ungureanu and Zografos n.d.).

However, this decentering shift leads to a terra incognita, characterized by conflicting interpretations of the Anthropocenic crisis of ecological degradations and societal disruptions (Tønder 2017; Bonneuil and Fressoz 2016). These interpretations can overlook the inherently dialectical relationship between humans and nature, which encompasses both a profound interdependence and care alongside inherent tensions, aporias, and power dynamics.[1] On one end of the spectrum, utopian visions abolish the complexities of the dialectic relationship highlighted by critical environmentalism by portraying a romanticized harmony between humans and idealized nature. On the other end, dystopian and catastrophist visions prominent in popular cinematic imagination exploit the emotional impact of climate anxiety and trauma and exalt the fatal clash between human beings and ecosystems.

In the Anthropocene as a time of crisis (Beistegui 2025), artistic imagination emerges as a privileged and pivotal force in reconsidering alternative

DOI: 10.4324/9781003468028-13

relationships between human beings and nature. This chapter analyzes the sci-fi film *Annihilation* (Alex Garland, United Kingdom/United States, 2018) against the background of Jeff VanderMeer's homonymous novel (2014).[2] Both works are examined as imaginative approaches that navigate human–nature relationships, transcending the conventional perception of sovereign subject and nature as an external object to be mastered or controlled (Iossifidis and Garforth 2022; Coby 2020). Garland's *Annihilation* follows the novel in chronicling the journey of a small team of expert women (as in the book, the protagonist is a biologist) whose mission is to enter and investigate Area X, or the Shimmer, a mysterious alien nature formed around a lighthouse and whose inexorable expansion threatens humanity (Deurwaarder 2018; Coby 2020; George 2018). Nonetheless, Garland's and VanderMeer's "brave new weird" (Ulstein 2017) constructions of nature are, we argue, ultimately divergent.

Firstly, the film juxtaposes three narrative threads and images of nature. In the film, the relationship between humans and alien nature is halfway portrayed as a Hall of Mirrors: everything is reflected in everything. There is neither a subject–object dichotomy nor a transparent barrier between species, organic and inorganic matter. The relationship human–nature is marked by "monstruous" hybridization, radical fluidity, and "vibrant matter" (Barad 2007; Bennett 2010).[3] Yet Garland's *Annihilation* only partially portrays this alternative radical vision; as the movie progresses, it intriguingly interweaves two additional narrative threads and images of the alien nature, leading to a Janus-faced finale. In the second narrative thread, the protagonist Lena (the biologist, played by Natalie Portman) seemingly triumphs over her alien doppelganger generated by the Shimmer, which results in the disappearance of the entire alien lifeform or ecosystem. This narrative resolution reduces the intricate dialectical issue of the ethical connection to nature to a mere psychological conflict within the individual protagonist's psyche. The final scene further complicates the movie's narrative by introducing ex machina a third narrative thread, the familiar science fiction trope of body snatching. This trope undermines the notion of hybrid, posthumanist relationships (Paszkiewicz et al. 2021) as it entails that the biologist-protagonist and notably her husband, Sergeant Kane (Oscar Isaac), are turned into tools of an ominous extra-terrestrial invasion. This inverted dystopian scenario positions humans as mere instruments of a malevolent alien nature.

Secondly, we argue that VanderMeer's *Annihilation*, by focusing on the biologist's transformation following her encounter with the Shimmer, consistently raises the question of transcending the rational ego and assimilating into a novel transhumanist[4] understanding of nature and existence. This evolved state moves past traditional dichotomies between subject and object, humans and environment, and challenges the perspective of nature as a mere entity to be controlled and harnessed in service of humanity. VanderMeer's narrative dismantles age-old dichotomies and portrays the "annihilation" of the rational ego as an affirmative and liberating process. In contrast to Garland's scenario, the "promise of monsters," to use Donna Haraway's phrase (2004), does not fall short, as it signifies a departure from limiting constructs and embraces a transformative liberation (Shildrick 2001; Asma 2009).

Finally, from the standpoint of critical environmentalism,[5] we contend that both artistic works only partially succeed in reimagining the human–environment relationship within the Anthropocene context. The regression to a traditional portrayal of an internal journey of the self, intertwined with a marital crisis (as seen in Garland's work), ultimately presents an anthropocentric scenario. Conversely, the total dissolution of the ego via bodysnatching (as depicted by Garland) or the emancipatory act of self-abandonment (as envisioned by VanderMeer) represent the dystopian and utopian facets of a regressive anti-humanist fantasy.

In the following, we develop a double-layered argument (hermeneutic and critical) by interpreting the movie as a sci-fi psychological thriller "undecided" between an anthropocentric and an alien invasion scenario. In turn, the novel centers on the experiential onto-ethical issue of the transformation of subjectivity in relation to the alien nature. We reconstruct the mission's premise and the events unfolding within the Shimmer, along with hypotheses surrounding the alien ecosystem that is both monstrous and sublime. Subsequently, we analyze the divergent resolutions of the film/literary work and their symbolic implications for a critical examination of the Anthropocenic condition.

The Encounter with the Shimmer: Fluid Nature and "Monstruous" Hybridization

Garland's science fiction narrative unfolds as a psychological thriller, skillfully weaving together scenes from different temporal junctures. These scenes include flashbacks delving into Lena's past, detailing her relationship with her husband, Sergeant Kane, and her infidelity with Dan. Concurrently, the film portrays the present timeline, following the unfolding mission into the Shimmer, alongside glimpses of flash-forwards depicting Lena's debriefing and her unsettling final reunion with the alienated Kane.

The movie commences by introducing the brilliant Lena, a Johns Hopkins biology professor, engaging her students in the realm of cell biology and cellular life. Lena specializes in the genetically programmed life cycle of a cell. In particular, she introduces students to the formation of tumor cells, announcing cancer as the semester's focal topic. This thematic prelude anticipates the alien expansion within the Shimmer gradually colonizing and destroying human life. However, Lena is portrayed as a science optimist, fostering a belief in the potentiality of overcoming cellular mortality by "circumventing the Hayflick limit."[6] In a flashback dialogue with Kane before the mission, he observes that "God does not make mistakes." Lena retorts: "Sure, He does. If you take a cell, circumvent Hayflick limit you can prevent senescence. Cells keep dividing, they do not die. They say aging is a natural process, but it's actually a fault in our genes."

However, Lena's life is deeply affected by a profound wound: the loss of her husband, Sergeant Kane, who vanished during a covert mission a year prior. Dan, her colleague, and former lover, encourages her to move forward. Unexpectedly, Kane reappears at their home, but he seems alienated and severely unwell. Kane's inexplicable return after so long culminates in his collapse, leading both Lena and Kane to be taken to a secret military facility. There, the biologist learns about Area X: three years prior, an enigmatic extra-terrestrial lifeform landed near a lighthouse, triggering the creation of a peculiar ecosystem. Since its arrival, this alien ecosystem has relentlessly expanded, posing an existential peril to humanity. Although her husband is the sole individual to have returned from Area X, he is in a deteriorating state (multiple organ failure, massive internal bleeding), facing imminent death (Deurwaarder 2018).

In Garland's psychologizing vision, Lena grapples with overwhelming guilt regarding her husband's condition and impending demise. She harbors a justified suspicion that Kane might have volunteered for the perilous mission into the Shimmer as a form of self-destructive intent, possibly influenced by her past involvement with Dan, with whom she had an affair. This suspicion adds to Lena's burden of guilt and emotional turmoil, attributing a sense of responsibility for her husband's tragic situation.[7] Driven by her sense of guilt, Lena is resolute in her determination to join the new expedition to Area X, propelled by the hope of unraveling its mysteries and potentially rescuing her husband. Unlike previous military missions, this new team comprises women scientists and a paramedic specifically selected by Dr. Ventress (Jennifer Jason Leigh), the team's leading psychologist. However, it becomes apparent that all five team members—Lena, Dr. Ventress, the paramedic Anya, the physicist Josie Radek, and the geomorphologist Casy Sheppard—are what Sheppard describes as "damaged goods." Dr. Ventress is grappling with terminal cancer, Sheppard mourns the loss of her young daughter to leukemia, while Josie and Anya have struggled with trauma or addiction. As it turns out, the addition of this psychological dimension in Garland's film is crucial, as the encounter with the alien lifeform becomes a metaphor of the confrontation with one's unconscious.

The team initiates the mission tasked to venture into the Shimmer, acquire data, and return. The Shimmer is portrayed in both the cinematic and literary versions as a 'magical forest.'[8] This recurring motif evokes associations with fairy-tales, initiation narratives, and horror stories, while also echoing the cinematic topos of the 'zone.'[9] In the film, the crew's initial encounter with the forest is both bizarre and transformative: they awaken to a startling realization of memory loss, unable to recollect anything from the initial three-four days. Their disorientation is amplified by the uselessness of their scientific tools in the Shimmer: compasses, phones, and radio transmitters. The alien nature appears to create a closed ecosystem: any attempt to communicate with the outside world proves futile (Deurwaarder 2018).

Within the Shimmer, sunlight undergoes refraction, dispersing into a spectrum of rainbow-like colors. The diverse array of rainbow hues illuminating

Figure 11.1 Entering the Shimmer, the "enchanted forest."

The enchanted forest, particularly when viewed alongside Garland's psychologically driven script, can be seen as a metaphor for the subconscious. Journeying through this forest becomes a metaphorical exploration of facing one's inner demons and traumas, as will be seen in the next section.

the forest imbues numerous scenes in the movie with a sense of enchantment and otherworldly beauty. Despite any disparities between Garland and VanderMeer,[10] the nature and alien life within the Shimmer are simultaneously stunningly beautiful and ominously monstrous. In Garland's film, upon reaching a submerged house in a swamp, Lena makes a startling discovery concerning the phenomenon of transspecies hybridization: a rainbow of flower colors, forms, and species with distinct genetic makeup inexplicably stemming from the same root. Similarly, evidence of transspecies hybridization surfaces when Lena discovers shark teeth in the mouth of a gigantic crocodile she manages to slay in a dramatic confrontation (Lena remarks, "it's a crossbreed, although you cannot crossbreed between different species"). The exuberance of the alien landscape manifests in simultaneously monstrous and wondrous forms: Garland skillfully crafts cinematic scenes reminiscent of what Edmund Burke termed "delightful horror"—such as when describing a tempest. In one such scene, the director presents enchanting, yet unsettling images of deer-like creatures and eerie, translucent forms observed at the lighthouse.

An aesthetic rooted in the sublime heavily influences VanderMeer and Garland's portrayal of nature, as elucidated by Burke (1998). This aesthetics merges the peculiar allure of the uncanny and otherworldly beauty with elements of horror, triggering contrasting emotional impulses of attraction and revulsion, awe and terror (Burke 1998).[11]

In both works, traditional understandings of nature are unsettled by uncanny processes of hybridization melting the human/nonhuman animal and animal/plant divides. In VanderMeer's literary work, the biologist marvels at dolphins while also experiencing a sense of unsettlement at the emergence of a "humanimal" (Segarra 2022): "As they (the dolphins) slid by, the

Figure 11.2 The "deer" with flowered horns.

nearest one rolled slightly to the side, and it stared at me with an eye that did not, in that brief flash, resemble a dolphin eye to me. It was painfully human, almost familiar" (VanderMeer 2014, 112).[12] Correspondingly, in Garland's film, the horror manifests in gruesome violence and an ever-more unsettling process of hybridization. Upon reaching an abandoned residence within the area, the crew encounters a monstrous bear- and human-like creature that viciously attacks them, resulting in Sheppard's demise. Shockingly, the hybrid creature (the "humanimal," in Segarra's terms, 2022) begins to mimic Sheppard's voice after her death.[13]

Figure 11.3 The confrontation with the monstrous "bear" using Sheppard's voice.

Figure 11.4 The hybrid human plants.

The chilling effect of hearing Sheppard's voice emanating from the monstrous bear, which also claims the life of Anya, is profoundly disturbing. As Coby reflects, it prompts a sense of discomfort generated by a nature that defies any expectations. It also induces a profound terror rooted in the realization that "our best efforts will not save us, not in any recognizable way" (2020, 16).

As they journey deeper into the Shimmer, Lena undergoes a process of "contamination." As if intimately penetrated by the alien nature or lifeform, her cells exhibit a vibrant array of rainbow-like colors and divide into various elements at a staggering pace, leading to permanent mutations. As a biologist, she is equally fascinated and horrified by the incomprehensible nature of her transformation. Furthermore, as the remaining squad reaches a deserted village, they encounter human-sized natural formations that appear to be the result of hybridization between humans and plants.

Lena and Josie have a crucial dialogue concerning this strange phenomenon of hybridization occurring in the Shimmer. Josie ventures an explanatory hypothesis concerning the "human plants": every leaf has the "human Hox genes." As the biologist clarifies, these are "genes that define the body plan, the physical structure." Josie confirms this view: "the plants have a human body plan." And she further asserts that "the Shimmer is a prism, but it refracts everything. Not just light and radio waves. Animal DNA, plant DNA, all DNA." In other words, within the Shimmer, nature operates through rapid 'gene jumps' and instantaneous genetic exchanges—specifically, through cross-species hybridization and dynamic intermingling of genetics. Humans are 'echoed' or mirrored into plants just as the plants are reflected into humans. As the biologist observes in VanderMeer's novel,

> I was convinced that when I wasn't looking at them, these cells became something else, that the very act of observation changed everything. I knew this was madness and yet still I thought it. I felt as if Area X was laughing at me then—every blade of grass, every stray insect, every drop of water. ... The wood was indeed wood. The rat was indeed a rat. The moss and the fox ... were composed of modified human cells.
>
> (2014, 159)

This representation of the alien nature turns upside down key elements of the traditional vision of nature in the Western tradition as a relatively fixed and hierarchical order made of separated 'kingdoms' and 'species.' In both the novel and the film, the essence of the mysterious ecosystem is hybrid and unconventional. It diverges from traditional pre-Darwinian static perceptions of nature and challenges the Darwinian concept of the tree of life, which posits gradual evolution through natural selection over time. Within the Shimmer, there exists no rigid identity, no clear-cut boundaries between species, and no distinct division between observer and observed. Changes and transformations are not driven by the gradual unfolding of natural selection across time. Instead, the core principle lies in genetic reflection, where everything mirrors everything else, fostering mutations and an unceasing flow of genetic information.[14]

If Anya and Sheppard are devoured by the monstrous Shimmer, Josie chooses a different path—that of acceptance. In a scene that is not drawn from the novel, she remarks to Lena: "Ventress wants to face it. You want to fight it. But I don't think I want either of these things." At this point, the camera reveals that plants are growing inside Josie's forearms. After this unsettling revelation, Josie abruptly leaves and seemingly fuses with the environment without further ado, accompanied by dreamy music and Lena's befuddled and terrorized look. Whereas this is only a passing scene in the movie, a somehow similar embracing of the new nature and dissolution of the ego is key in the novel (as we will discuss).

The End or Triumph of Nature as Alterity? Diverging Scenarios

Between Cinematic Psychodrama and Bodysnatching

As the movie and the book reach their peak and denouement, the two visions take different paths. In Garland's cinematic interpretation, the surviving Ventress and Lena reach the lighthouse, believed to harbor crucial insights into the Shimmer. Both in the movie and the novel, the lighthouse is depicted as "vibrant matter" (Barad 2007; Bennett 2010), presenting itself as a strange and mutable life form imbued with agency, adapting and merging with humans upon contact.

In the movie, though, once inside the lighthouse, Lena finds a video camera and watches footage seemingly made by Kane. As it turns out, the older Kane is filmed by his identical double.[15] Kane speaks in front of the camera: "I thought I was a man. I had a life … And now I am not sure … Was I you? Were you me? My flesh moves like liquid. My mind is cut loose. I can't bear it." Surprisingly, he blows himself up. We see the "other" Kane coming from behind the camera. This is shocking and new: so far, the Shimmer has been working through hybridization. In the lighthouse, the original site of the Shimmer, the impact of the alien lifeform is different: it does not act only through hybridization but also through doubling. Does this scene represent Kane's

Figure 11.5 Lena struggling with her mimetic double generated by the Shimmer.

metaphorical confrontation with his traumatic past, and his separation from the older life and I? Or is he annihilated or absorbed by the new alien reality? The answers to these questions remain suspended.

Inside the lighthouse, there is a hole that leads into a dark cave of sorts that is interpretable as a spatial metaphor of the Id. Lena enters the hole and encounters Ventress, who seems possessed: "Unfathomable mind … It's inside me now. It is unlike us. I don't know what it wants. Or if it wants. But it will grow … Annihilation." Immediately afterward, Ventress is annihilated by the Shimmer: a thick and violent beam of light comes out through her mouth; colorful cell formations float and divide at an unnatural speed.

In the next scene, the Shimmer creates Lena's duplicate, out of a drop of blood, who imitates her moves. In a long, tense suite of shots, she struggles with her double and emerges victorious. As a result of this struggle, the whole Shimmer disappears suddenly and without explanation. Lena is free to return to the base.

The movie ends with the biologist's short debriefing and her encounter with Kane:

Interrogator:	So it was an alien …. Can you describe its form?
Lena:	No … It mirrored me … I attacked it … It was not destroying … It was changing everything. It was making something new.
Interrogator:	Making what?
Lena:	I don't know.

Whereas this dialogue points to a novel and non-destructive alterity, the last scene of the encounter with Kane gives the story a different twist. In a low-lit, cold, and tense atmosphere, the two exchange a few words:

Lena: You aren't Kane, are you?
Kane: I don't think so.
Kane: Are you Lena?

She does not answer. Instead, they hug. We see first a close-up on his face: the expression of his eyes is alienated and evil. Then, the camera lingers on her face. She appears to be herself, yet a subtle shimmer seems to shine in her eyes. In a first interpretation, the two have become aliens and completely abandoned their humanity: the extra-terrestrial nature has taken human form to spread and conquer. This body-snatching hypothesis seems to be confirmed, first and foremost, by the evil intentionality expressed by his last look. While this latter reading has a certain plausibility, it is not supported by the development of the movie. By envisioning the alien-nature formation as pure evil and malignantly expansive, nature's alterity is reduced to a cliché of alterity. This portrayal deviates from the previously enigmatic nature of the Shimmer to conform to a sci-fi stereotype, thus precluding the fruitful imagination of posthuman kinships and dilemmas.

In a second psychologizing interpretation, the focus is on Lena and her marital crisis, which is a crucial theme throughout the film that is absent from the book. The female biologist wins the encounter with the 'alien' nature, inner monsters, or double. Driven by guilt, she saves Kane, but he is now estranged from her: in this predominantly anthropocentric interpretation, their love and complicity have vanished, and they have become alien to each other. In the final analysis, the finale remains ambivalent: the marital story and the body-snatching motif seem to be juxtaposed rather than integrated organically in a coherent vision.[16] The movie's conclusion remains Janus-faced in a way that dilutes the enigmatic premise of the film, reducing the question of nature and its alterity either to an anthropocentric scenario or to a sci-fi stereotype.

... back to the onto-ethical problematic

In contrast to the film, in the literary work, there is no depiction of any marital discord or delving into the psychological drama between the biologist and her husband.[17] VanderMeer's literary vision systematically avoids dealing with the psychological aspects of the biologist's journey and her team's exploration of the Shimmer. Narrated from a first-person perspective, the book sidesteps the examination of the characters' psychological traumas, their growth, or their affective interrelationships. Tellingly, the protagonists are nameless and devoid of personal histories or past traumas. While the film extensively incorporates direct dialogue and dynamic affective interactions, employing visually saturated shots in low light, predominantly green- or grey-toned, to evoke a sense of heightened emotional drama and inevitability, the book maintains a tone of introspection and tranquility throughout, portraying the biologist's transformation in a detached manner. Its primary focus lies in delving into the experiential and ontological shifts experienced by the biologist as a scientist, specifically highlighting her problematization of rationality and objectivity when confronted with the alien lifeform and its anomalous biological mechanisms. Scientific rationality and tools become as meaningless as the traditional categories based on the rigid dichotomizations of subject/object, organic/inorganic matter, autonomous I/environment, culture/nature. The matter itself is "vibrating" and seems endowed with agency: in the lighthouse, writing in English emerges on the walls in a way that is not human but is partly comprehensible. The tissue of the letters, of this sort of monstrous hybrid organism, only seems human:

> I kept squinting through the microscope lens, and raising my head, and squinting again, as if I couldn't see the sample correctly. … Was it really human? Was it pretending to be human? As I said, there were irregularities. … Which meant it couldn't be brain tissue, which meant it was definitely not human. I felt unmoored, drifting, once again.
>
> (VanderMeer 2014, 73)

The biologist puzzles over and gradually accepts this new reality that radically penetrates and transcends her 'I' and 'scientific rationality.' The only possible escape is to think differently, to understand beyond the individual self. Unlike the tumor cells that she is fascinated with as a university professor, the multiplying cells of the alien nature or lifeform do not bring about her annihilation as physical death, but the annihilation of a way of life, a dissolution of nature separated from culture, a dissolution of a certain humanism. For the biologist, the 'real' annihilation results from 'changing sides':

> The terrible thing, the thought I cannot dislodge after all I have seen, is that I can no longer say with conviction that this is a bad thing. Not when looking at the pristine nature of Area X and then the world beyond, which we have altered so much. Before she died, the psychologist said I had changed, and I think she meant I had changed sides. … If I don't have real answers, it is because we still don't know what questions to ask. Our instruments are useless, our methodology broken, our motivations selfish.
>
> (VanderMeer 2014, 192–3)

Annihilation is a new beginning—that of a transhuman life that may seem to fulfill the biologist's dream of overcoming cell mortality in Garland's vision. With the dissolution of the self and the transformation of her body, the protagonist completely supersedes humanity. Now she feels that she belongs there, in Area X:

> Observing all of this has quelled the last ashes of the burning compulsion I had to know everything… anything… and in its place remains the

> knowledge that the brightness is not done with me. It is just beginning, and the thought of continually doing harm to myself to remain human, seems somehow pathetic.
>
> (VanderMeer 2014, 194)

The biologist's fate is echoed by Josie's 'choice' in Garland's cinematic interpretation; however, Josie's acceptance is not exhilarating. It is portrayed more as an anguishing self-abandonment. Her gesture of self-annihilation and fusion with a monstrous, uncanny reality echo, and is interpretable, as being devoured by her inner monsters. Notably, Lena's reaction to Josie's departure is one of disconcerting and fearful anxiety, rather than of the celebration of a willing and self-embracing transition into a new life. On the contrary, in the novel there is a sense of exhilaration and purpose in the biologist's embrace of the new transhuman or nonhuman life: for VanderMeer, annihilation is affirmative.

Conclusion

The contemporary anthropocenic condition calls for a critical environmentalism built upon a profound restructuring of human–nature relationships and the overcoming of the dystopia/utopia binarism.[18] The displacement of anthropocentrism is marked by a 'double bind'[19] or open dialectic made of conflictive pulls, imperatives, and difficult negotiations. From the standpoint of a critical dialectical environmentalism, socio-political relations are deeply embedded within relations of interdependence with a damaged environment, prompting a call for a transition toward a politics centered on fostering care for the natural world and limiting the objectification of nature. Yet, these relationships are also characterized by complex tensions and disruptions generated by the exceptional environmental costs of sustaining a rapidly expanding human community. Both Garland's film and VanderMeer's novel draw on biological imagination to envision inspiring alternatives for overcoming the anthropocentric view of nature as an external object to be mastered (Iossifidis and Garforth 2022). Their biological fantasy replaces the slowly paced natural selection reactive to environment with the vision of hybridization amongst species, accelerated 'gene jumps,' and a vision of "vibrant matter" (Barad 2007; Bennett 2010). The human–environment relationships in the Shimmer are marked by radical hybridity and interconnectedness transcending the subject–object binarism and emphasizing nonhuman agency. This artistic imaginative portrayal is interpretable as a hyperbolic metaphor that conveys the awareness of a new relationship to the environment in the Anthropocene. It presents nature not merely as a 'commodity' or an idealized, perfectly harmonious entity but as a decentering otherness—depicted as monstrous or otherworldly beautiful, or even both simultaneously. In this perspective, humans are neither at the center nor detached from nature, exerting dominance over it. Instead, they are integral components within a network of dynamic relations involving hybrid and interacting elements.

However, Garland's film intertwines, as we have argued, other narrative threads in a contradictory fashion. Specifically, toward the movie's conclusion, Garland introduces two conflicting scenarios and portrayals of nature. The entire Shimmer, or the alien environment, appears to disintegrate due to the protagonist's symbolic triumph over her doppelganger (signifying the inner monsters) (Asma 2009). In this scenario, the eco-cinematic vision (Paszkiewicz 2023) is reduced to a way of exploring the crisis of Lena's ego and her marriage. The question of nature as alterity and onto-ethical problem is, thus, dissolved into a psychologizing anthropocentric scenario. Furthermore, the concluding scenes introduce ex machina, an element—a well-trodden science fiction trope involving body-snatching—that dismantles the potential for posthuman kinships. Lena and others appear to become mere instruments of an alien invasion and the subsequent conquest of Earth.[20]

Overall, in Garland's vision, the ultimate significance of the Shimmer remains as undecidable as Lena's role: is she the savior of Kane and humanity, or is she the mask of an (evil) alien agent? The resolution hangs in suspension, yet the portrayal of nature as malign remains, in this latter scenario, built on anthropocentric premises as well: the alien nature is understood as malign because it is malign *to us*.

In contrast, VanderMeer's *Annihilation* consistently poses the question of superseding the ego and integrating into a new life and reality. In the novel, an equally creative and enigmatic transhuman lifeform transcends age-old dichotomies between subject and object, humans and environment, as well as the vision of nature as an external object to be mastered and put in the service of humans.

Despite the differences between the literary and the cinematic works, the return to the conventional anthropocentric perspective of the dominating ego (Garland) and, conversely, the projection of its total dissolution (Garland/VanderMeer), merely reflect one another like a mirror image. Both scenarios abolish the dialectical nature of the relationships between humans and the environment and take refuge either in a dystopian fantasy (the alien nature is evil) or ecological obscurantism (the utopia of a total superseding and, thus, negation of human life).[21] These scenarios are equally unappealing when considering a critical reassessment of the transformation of subjectivity and the cross-pressures of the relationships between society and nature during the Anthropocene.

As the biologist reflects, "our instruments are useless, our methodology broken, our motivations selfish" (VanderMeer 2014, 192–3). Area X, then, defies scientific and rational epistemologies, instead imposing "a process of un-knowing" (Sperling 2019, 23–24) upon those who venture into its depths. *Annihilation* denies its readers a fixed interpretation (Luckhurst 2017). Both the novel and Area X itself unveil "incongruous experiences," leaving judgment in a state of suspension (Luckhurst 2017, 1045). Yet scientific instruments and methodologies are not problematic per se, and their abolishment is no reason for celebration;[22] problematic is the subordination of scientific research to a socio-political practice based on an extractivist approach to nature. While humans do not hold a central or privileged position within the vastness of nature, they have a singular potential to provide creative solutions to environmental problems. This potential is rooted in scientific research and requires a radical reorientation of human–environment relationships, drawing critically from diverse sources, including artistic imagination.[23]

Notes

1 Consider the environmental costs incurred by scientific research, green transitions, and agriculture (even of green agriculture) necessary for an ever-expanding human population (see, for instance, Wapner 2008; Pellizzoni, Leonardi, and Asara 2022).

2 This is the first part of *The Southern Reach Trilogy* (VanderMeer 2014). Our focus is the film against its literary background. The analysis of the whole trilogy is beyond our scope (Deurwaarder 2018; Coby 2020; Geroge 2018; Carroll 2015; Dowd 2018; Ulstein 2017).

3 Here we extrapolate the concept of "vibrant matter" as articulated in Karen Barad and Jane Bennett's significant work (Barad 2007; Bennett 2010). Barad and Bennett argue that matter is not passive or inert but rather vibrant and lively, constantly entangled with the world around it. Moreover, they suggest that matter is not simply something that exists independent of human observation but is instead an active participant in the ongoing processes of the universe. Barad's and Bennett's notion of vibrant matter challenges traditional dualisms between subject and object, human and nonhuman, and emphasizes the interconnectedness and agency of nonhuman entities as well. In anticipation, in *Annihilation*, the conventional inorganic matter (such as a structure like the lighthouse) takes on organic characteristics akin to an agency-endowed living entity in both its literary and cinematic renditions.

4 In this context, we use "posthumanist" and "transhumanist" in a specific way. "Posthumanist" refers to de-centered ways of conceiving the relationship between human and nonhuman (see also Braidotti 2013). "Transhumanist" specifically refers to radically transcending the human condition toward a completely different nonhuman reality. We are not gesturing toward a transhumanism that is often used with reference to a cultural-philosophical theory that advocates for the use of science and technology to enhance human physical and cognitive abilities beyond what is considered typical or "normal."

5 We take critical environmentalism as avoiding both the utopian view of a romanticized harmony between humans and an idealized nature and the dystopian one of destructive conflict between the two. A critical environmental view takes distance from these opposite scenarios and recognizes the complexities of humans and nature, and the double relationship of care and interdependence, on the one hand, and tension and conflict, on the other (Wapner 2008; Brincat and Gerber 2015; Pellizzoni, Leonardi, and Asara 2022).

6 More specifically, Lena holds the belief that senescence and death are errors in God's design, suggesting that with the aid of modern biological knowledge and technology, it is possible to overcome these limitations. She proposes the idea of "circumventing the Hayflick limit" as a means of correcting this perceived mistake. What this entails is bypassing the Hayflick limit, a phenomenon where most differentiated cells in our body have a finite number of divisions. Beyond a certain

limit, these cells enter senescence, halting further division. This limit is closely linked to changes within chromosomes as cells age. Chromosomes possess special structures at their ends, known as telomeres. With each cell division, telomeres shorten, and once they reach a critical length, cells cease to divide and undergo cell death. This process plays a crucial role in cellular aging and eventual death.

7 Dr. Vendress confirms this suspicion later in the movie.

8 Garland explains: "And the most defining thing about [the novel] was its atmosphere; the specifics of plot were not as crucial. Tone was very important. I didn't reread the book, which was a dreamlike, slightly hallucinatory novel. I did an adaptation of my memory of the novel, a slightly odd conceit" (Garland in Thompson 2018).

9 The locus classicus of this motif in cinema is Andrei Tarkovsky's sci-fi fantasy *The Stalker* (1978).

10 See also VanderMeer (2008) on the "new weird" and Iossifidis and Garforth (2022), Dowd (2018) and Luckhurst (2017).

11 By drawing freely on Edmund Burke's *Philosophical Enquiry into the Origin of our Ideas of the Sublime and the Beautiful* (1998), we distinguish two aesthetic forms of "sublime" that surface in Garland's film, that is, horror and otherworldly beauty, both to be found in *Annihilation*. Garland's poetics of monstrosity generating revulsion and fascination is achieved by juxtaposing images of horror and sublime beauty to recover the alterity of nature. Yet, there is a significant difference in tone between Garland and VanderMeer: a sense of horror and the ominous predominates in the cinematic work; while the literary work is more detached, preparing the biologist's transhuman transformation.

12 Upon uncovering Lena's hidden marriage to Kane, Anya feels betrayed. In a fit of anger, she restrains Ventress and Lena. In a horrific scene, the hybrid creature, emitting Sheppard's distressing pleas for help, consumes her.

13 In a psychologizing interpretation that is condoned by a good part of Garland's filmic interpretation, the journey in the magic forest stands for a confrontation with trauma. Accordingly, Sheppard is devoured by her trauma or inner daemons.

14 Cheng-Hin observes that "The Shimmer … transforms just for the sake of transformation. It is what Deleuze and Guattari would describe as a total field of deterritorialization: within its changing borders all established meanings are erased" (Cheng-Hin 2018, 2). In our view, Cheng-Hin's argument finds no support in either the movie or the book. The rationale behind the transformation being merely for the sake of change remains unclear. The conclusion of the movie implies that the transformation serves a purpose of conquest, diverging from the novel's script (as discussed later).

15 The motif of the monstruous doppelgangers is entirely introduced by Garland.

16 In a third hypothesis, the two represent a 'new humanity,' resulting from the hybridization between the alien reality and the humans. However, the last scene does not support this hypothesis: the Shimmer either completely disappears or body-snatches Lena and Kane.

17 Whereas the apex of the lighthouse scenes is, in the movie, the struggle with the double, in the novel the biologist 'confronts herself' only going through her husband's diary.

18 See also Ungureanu (2023).

19 We borrow this phrase from Jacques Derrida without its *specific* aporetic implications.

20 This end-of-the-world anti-humanist horror phantasy is the mirror image of traditional humanism. However, despite its cliché nature, it is not devoid of ambivalence, as it can be interpreted as radically anti-anthropocentric, given that it eliminates the human from the scenario.

21 The complete abandonment of rational autonomy is a regressive fantasy that verges on ecological obscurantism. Equally, the dissolution of the ego resonates with valuable spiritual-mystical traditions such as Buddhism.
22 This is Iossifidis' and Garforth's position (2022).
23 For an empirical approach to the impact of VanderMeer's *Annihilation*, see Iossifidis and Garforth (2022).
Finally, the authors would like to thank the editors of this volume for their critical suggestions on a previous version of this chapter. Any remaining errors are our own.

Works Cited

Asma, Stephen T. 2009. *On Monsters: An Unnatural History of Our Worst Fears*. Oxford: Oxford University Press.

Barad, Karen. 2007. *Meeting the Universe Halfway: Quantum Physics and the Entanglement of Matter and Meaning*. Durham: Duke University Press.

Bee, Julia. 2020. "Filming through the Milieu. Becoming Extinct and the Anthropocene." In *The Anthropocenic Turn. The Interplay between Disciplinary and Interdisciplinary Responses to a New Age*, edited by Gabriele Dürbeck and Philip Hüpkes, 191–211. New York: Routledge.

Beistegui, Miguel. 2025 (forthcoming). *A Philosophy of Crisis*. Chicago: The University of Chicago Press.

Bennett, Jane. 2010. *Vibrant Matter*. Durham: Duke University Press.

Bonneuil, Christophe, and Jean-Baptiste Fressoz. 2016. *The Shock of the Anthropocene: The Earth, History and Us*. Translated by David Fernbach. London: Verso.

Braidotti, Rosi. 2013. *The Posthuman*. London: Polity.

Brincat, Shannon, and Damian Gerber. 2015. "The Necessity of Dialectical Naturalism: Marcuse, Bookchin, and Dialectics in the Midst of Ecological Crises." *Antipode* 47 (4): 871–893.

Burke, Edmund. 1998. *A Philosophical Enquiry into the Sublime and Beautiful*. London: Penguin Classics.

Carroll, Siobhan. 2015. "The Ecological Uncanny: On the 'Southern Reach' Trilogy." Reeview of *Annihilation*, by Jeff VanderMeer. *Los Angeles Review of Books*, October 5, 2015.

Cheng-Hin Lim, Alvin. 2018. "*Annihilation* and the Modes of Becoming." ippreview.com/index.php/Blog/single/id/674.html.

Clot Garell, Anna. 2024. "Thinking through Ruination: An Introduction." *Special section: "Thinking through ruination: theoretical and empirical approaches to the ruins of the Anthropocene"*, edited by Anna Clot Garrell. *Digithum* 31. https://doi.org/10.7238/d.v0i31.426745.

Coby, Jim. 2020. "Reading Jeff VanderMeer's *Annihilation* in the Anthropocene." *Journal of Science Fiction* 4 (1): 15–16.

Deurwaarder, Beau. 2018. "Nothing Comes Back: *Annihilation* as a Posthuman and Anthropocene." Paper presented at the *Australasian Society for Continental Philosophy (ASCP) Conference at Western Sydney University*, Parramatta, November 23, 2018.

Dowd, A.A. 2018. "Alex Garland's *Annihilation* is weirder, wilder, and much scarier than *Ex Machina*." *A.V.Club*, February 22, 2018. https://www.avclub.com/alex-garlands-annihilation-is-weirder-wilder-and-much-1823233848.

Garland, Alex. 2018. "*Annihilation* Director Alex Garland Speaks Out on Screwing with Genre and Studio Panic Attacks." *Indiewire*, April 8, 2018. https://www.indiewire.com/features/general/annihilationdirectoralex-garland-devs-television-1201950154/.

George, Joe. 2018. "The Philosophy of Self-Destruction in Alex Garland's *Annihilation*." *Tor.com*, June 20, 2018. https://www.tor.com/2018/06/20/the-philosophy-of-self-destruction-in-alex-garlands-annihilation/.

González-Hidalgo, María, and Zografos, Christos. 2020. "Emotions, Power, and Environmental Conflict: Expanding the 'Emotional Turn' in Political Ecology." *Progress in Human Geography* 44 (2): 235–255.

Haraway, Donna. 2004. "The Promises of Monsters." *The Haraway Reader*, 63–124. New York: Routledge.

Iossifidis, Miranda Jeanne Marie, and Lisa Garforth. 2022. "Reimagining Climate Futures: Reading *Annihilation*." *Geoforum* 137: 248–257.

Luckhurst, Roger. 2017. "The Weird: A Dis/Orientation." *Textual Practice* 31 (6): 1041–1061.

Paszkiewicz, Katarzyna. 2023. "Facing the End of the World: *Take Shelter* as Horror Ecocinema." *Journal of Film and Video* 75 (3): 22–37.

Paszkiewicz, Katarzyna, Andrea Ruthven, and Marta Segarra. 2021. "Posthuman Kinship." *Dictionary of Gender in Translation*. https://worldgender.cnrs.fr/en/entries/posthuman-kinship/.

Pellizzoni, Luigi, Emanuelle Leonardi, and Viviana Asara, eds. 2022. *Handbook of Critical Environmental Politics*. Cheltenham: Edward Elgar Publishing.

Segarra, Marta. 2022. *Humanimales. Abrir las fronteras de lo humano*. Barcelona: Galaxia Gutenberg.

Shildrick, Margrit. 2001. *Embodying the Monster: Encounters with the Vulnerable Self*. London: Sage.

Sperling, Alison. 2019. "On Not Returning Home, an Introduction." Alison Sperling Institute for Cultural Inquiry. *SFRA Review* 330 (Fall): 23–39.

Thompson, Anne. 2018. "'Annihilation' Director Alex Garland Speaks out on Screwing with Genre and Studio Panic Attacks." *Indiewire*, April 8, 2018. https://www.indiewire.com/2018/04/annihilation-director-alex-garland-devs-television-1201950154/.

Tønder, Lars. 2017. "Five Theses for Political Theory in the Anthropocene." *Theory & Event* 20 (1): 129–136.

Ulstein, Gry. 2017. "Brave New Weird: Anthropocene Monsters in Jeff VanderMeer's *The Southern Reach Concentric*." *Literary and Cultural Studies* (March): 71–96. DOI: 10.6240/concentric.lit.2017.43.1.05.

Ungureanu, Camil. 2023. "Socialist Cinema, Gender, and the Robot Revolution: Beyond Utopia and Dystopia." *Studies in European Cinema* 4 (1): 44–60.

Ungureanu, Camil, and Christos Zografos. (n.d.). "Green Transition, Sacrifice, and Crisis of Masculinity: *Alcarràs* and *The Beasts* between Nostalgia and Resistance."

VanderMeer, Jeff. 2008. "Introduction. The New Weird." In *The New Weird*, edited by Ann VanderMeer and Jeff VanderMeer, ix–xvii. San Francisco: Tachyon Publications.

VanderMeer, Jeff. 2014. *Annihilation*. London: Fourth Estate.

Wapner, Paul. 2008. "The Importance of Critical Environmental Studies in the New Environmentalism." *Global Environmental Politics* 8 (1): 6–13.

12 Inhabiting a Viral Culture

Verena Andermatt Conley

"My video went viral on YouTube!" It is most teenagers' dream to post such a tweet. Eureka! The words shine in the aura of success. They tell us, too, that the teenager's excitement may be a symptom of the hype and hoopla circulating at the speed of light in our contemporary culture. COVID-19 notwithstanding, it is little wonder that "viral" crisply describes how news disseminates and diffuses in the digital age, where speed and hyper-connectedness are the measure of all things. In addition to the metaphoric use of "viral" and leading up to the pandemic of COVID-19, over the past decade, a plethora of horror films and disease thrillers have foretold or programmed the devastations we are witnessing. Telling stories of how viruses infiltrate, infect, and kill humans, like COVID-19, they have seemed to proliferate. Among others: *Outbreak* (Wolfgang Petersen, United States, 1995), *28 Days Later* (Danny Boyle, United Kingdom, 2002), *Carriers* (Àlex Pastor and David Pastor, United States, 2009), *Contagion* (Steven Soderbergh, United States, 2011), *Containment* (The CW, United States, 2016), and *Bird Box* (Susanne Bier, United States, 2018). For the most part blockbusters, these and other films objectify a deadly virus by following well-known plot lines drawn between good and evil. Among them, *Contagion* stands out by reason of its measure, its ethos, and its treatment of what it means to inhabit our populous world. Directed by Steven Soderbergh, heralding an international all-star cast—Gwyneth Paltrow, Matt Damon, Kate Winslet, Laurence Fishburne, Marion Cotillard, Jude Law, Brian Cranston, and the list goes on—when first released the film attracted little attention. Since the start of the scourge, however, and widely distributed, it began streaming on Amazon Prime, Hulu and other popular services. If it hasn't exactly gone "viral," the feature has received much wider acclaim during the spread of a pandemic that a decade ago it foresaw with uncanny accuracy.

Contagion is about a sudden outbreak on a global scale of a yet unknown, deadly virus said to follow SARS-CoV (2003) and H1N1 (2009), except that, unlike COVID-19, it indiscriminately strikes old and young. The film begins on what we are told is "Day 2," at Chicago's O'Hare airport. A black screen accompanies indistinct airport noises. In the darkness we hear a woman coughing. When the image track fades in, we see Beth Emhoff (Gwyneth

DOI: 10.4324/9781003468028-14

Paltrow), an attractive American business executive, sitting at a counter of an airport bar, contemplating a glass of beer and a bowl with peanuts. She coughs again and, so it seems, experiences a malaise. Her cell phone rings. When she answers, we hear the voice of a man who identifies himself as John Neal (Robert G. Beck), presumably a former lover. He laughingly chides Beth for leaving him without saying goodbye after having sex with him in a nearby hotel that afternoon. The scene smacks of adultery: he thanks her for calling him during her layover in Chicago, while on route from Hong Kong to Minneapolis. He urges her to contact him again—but only by using a special email address. Their conversation is interrupted when a loudspeaker announces the imminent departure of Beth's flight. After some quick cross-cutting, we see Beth arriving in a stately house in an affluent Minneapolis neighborhood where her loving husband, Mitch Emhoff (Matt Damon), and her young son welcome her. The following morning during breakfast, Beth falls to the kitchen floor. She is rushed to the hospital where—less than ten minutes into the film—she has a seizure and dies. Within hours, her son, also stricken, is the next to die. At first Beth's malaise, apparently a punishment for her adulterous behavior, seems a strand of the moral thread of the narrative, which ultimately is left in suspension.

The film soon cross-cuts to other global hubs, to shots of Hong Kong, London, Tokyo, San Francisco, and the Guangdong Province in southeast coastal China. It then turns to interior spaces, where emergency meetings are held at the Center of Disease Control and Prevention (CDC) in Atlanta and the World Health Organization (WHO) in Geneva. The two major agencies begin to grapple with the emerging pandemic, spreading everywhere, "going viral," as it were. The urban centers are identified in intertitles in red or white letters enumerating the names of the cities, the size of their population, and the date. The quick succession of facts and figures gives a sense of the speed at which the virus is spreading as well as a growing sense of dread. In each urban area, except for Atlanta and Geneva, at least one character is stricken and dies. The different locations do not so much situate the singular stories than link them through a series of what some time ago anthropologist Marc Augé had called *non-lieux*, non-places—airports, planes, subways, taxis, hotels—or places that could be "anywhere whatsoever" and whose blandness suggests a hyper-connectivity of people from all corners of the globe. The sheer number of toponyms mentioned underlines the high risk of a rapid and worldwide transmission of the deadly virus.

After Day 3 (that is, following the death of Beth Emhoff and the emergency meetings at the CDC and WHO that ponder how to contain the virus while tracing its origin), the narrative is told as a race to develop a vaccine and an investigation that unfolds in a careful and well-timed series of flashbacks. Seeking desperately to reach the first person infected, the organizations track down the activity of those who fell victim to the virus earliest. By way of the CDC (under the direction of Dr. Ellis Cheever [Lawrence Fishburne] and an epidemic intelligence service officer, Dr. Erin Meers [Kate

Winslet] as well as WHO's Dr. Leonora Orantes [Marion Cotillard]), the film follows multiple leads that ultimately converge upon Beth Emhoff. Spread over 133 days, making use of television broadcasts, teleconferencing, security footage, and a panoply of digital devices, the investigation has agents progressively track down Beth's activities and trace her contacts until, at the end, in the very last shot, labeled Day 1, the film reveals the origin and beginning of the pandemic: on the evening before her return to Minneapolis, Beth is shown gambling in a casino at Macau, that is, in another non-place emblematic of what culture critics used to refer to as the "great casino society," a microcosm of simulation, another *non-lieu* or atopia defined by absence of borders as well as any symbolic or meaningful exchange (Baudrillard 1983; Augé 2008). On that last evening, documented mainly through pictures from security footage and from her own smartphone. Beth is surrounded by several characters who, we learn, all work for AIMM Alderson, an international mining and manufacturing company for which, fittingly, she is a marketing operations manager.[1]

Several casino workers, in addition to the characters who surround Beth in the pictures taken during a dinner to celebrate the construction of a new company factory, are those we had seen earlier, dying in buses, planes, airports, taxis and hotel rooms all over the globe. We also catch a glimpse of the now-stricken lover from Chicago who is carried out of his house on a stretcher while a disconsolate, hand-wringing wife stands by, regretfully informing the medical personnel that she and her husband both suddenly felt ill that morning. Dispatched by the CDC, Dr. Erin Meers (Winslet), who is heroically investigating the origins of the virus, also soon dies in the line of duty. Dr. Leonora Orantes (Cotillard), envoy of WHO, after being dispatched to Hong Kong and Macau, is taken hostage by a Chinese villager working at the casino who wants to get hold of the vaccine to help his impoverished village that, he fears, will not have access to the drug reserved for a wealthier clientele. Cotillard is released, thinking that she has successfully negotiated the transfer of the vaccine to the villagers. When at the airport on her way home she realizes that they had been duped and that what had been given to the villagers in exchange for her release was not a vaccine but a placebo, she abruptly leaves, presumably to return to the village and side with its inhabitants.

Contagion's plotline forecasts with disquieting accuracy much of what we have been witnessing in the continuous surge of COVID-19: fear and panic, a race to trace the origin of the virus, to develop a vaccine, individual heroism—but also intrigues among agencies and private citizens, widespread corruption amidst general misinformation. The film even features a blogger, aptly named Alan Krumweide (Jude Law) who, besides leaking the misconduct of the director of the CDC for breaking secrecy when he urges his wife to flee Chicago before a lockdown, also makes money roguishly by marketing forsythias as a cure. (The hoax blueprints former President Donald Trump's advice that Americans cure themselves of COVID by swallowing repeated doses of hydroxychloroquine.) The pace of the movie, the rapid montage, the frenzied movements of the characters, and the interwoven threads of the plot and the subplots add to the feeling, noted above, of dread pervading a viral, global culture calibrated at high speed and given over to a logic of hyper-connectivity.

Unfolding in a flat tone, without passing moral judgment, the film is implicitly and at times explicitly critical of a contemporary ethos dictating ways we inhabit our milieus in what it depicts as a viral culture of speed, that is, a culture based on glitter, money, consumption, and greed—in short, devoid of symbolic exchange, a world oblivious to complexity. On Day 133, when the distribution of vaccines at last takes place, set in the narrative just before the final revelation of Day 1, the film inserts a short sequence that leads up to the contamination. Undated, it is not shown as a flashback, but as something that both is and is not part of the story, something more general that repeats itself. The sequence depicts heavy machinery, bearing the logo AIMM Alderson, whose buzz saws and skidders wantonly fell acres of tall palm trees, presumably for the construction of new farms in Guangdong Province. The clear-cutting displaces a flock of bats, one of which (at least for the plot of *Contagion*) is followed in its migration to a banana tree where it picks on a fruit, and then to a pig farm where, hanging from the rafters, it drops a small piece of its bounty that lands on the floor of the pigsty. As soon as a piglet eats it, its own virus mixes with a bat virus. Hoisted onto a truck with others, it is presumably transported to a live animal market. Its fate is shown to emphasize our inhumane and loathsome treatment of animals. It turns up on the dinner table in the very casino of Macau where Beth spends her last celebratory evening. It is there, on Day 1, shaking the unwashed hands of the chef who had prepared the piglet for dinner, that she is contaminated. Here and elsewhere, the film avoids explicit moral judgment, be it that of bad hygiene, the general destruction of the planet through expansionism, or the demise of personal relations. We simply see an emotional husband looking at pictures of his wife in sartorial elegance, taken during the fancy dinner that had concluded with the fatal handshake.

From a broader perspective, however, by showing more than telling, the film impugns an ethos or a way of inhabiting that defines Beth's fast-paced, glitzy lifestyle, a condition we can assume endemic to a viral culture of global proportion. The film reveals Mitch's pain when he grasps Beth's smartphone before taking a picture of his teenage daughter and her now-vaccinated partner, who dance in the living room below. He looks at the images of her dinner at Macau while, from the living room, we hear the voice of Bono of U2 that the daughter had chosen to stream, singing, "You say you want diamonds on a ring of gold ... when all I want is you." Mitch tears up upon hearing the song while looking at pictures displaying the splendor and luxury his dead wife had craved. Temporarily unemployed, Mitch is a stay-at-home husband. He moves at a slower pace and occupies a different space than his wife who, living in the atopia or placelessness of a viral culture, was gambling with her life. The film makes an implicit critique of a culture whose ethos can turn deadly. If it shows contamination occurring through what the spectator can interpret as a lack of hygiene, and if it emphasizes the dangers of touching—people, credit cards, jars with peanuts, glasses, elevator buttons—*Contagion*, with the insert that registers the felling of palm trees, also lays stress on a global corporation's malfeasance, its short-sightedness, and its negligence. It thoughtlessly razes a forest crucial for the well-being of the planet and thereby helps foster a pandemic through a sheer lack of concern or foresight.

Perceived upon its release as yet another film capitalizing on dread and fear, *Contagion* spells out, in hindsight, the ethos at the origins of recent and current pandemics: deforestation, industrial agriculture, lack of responsibility on the part of global megaliths, a sizable increase of the human population that all lead to indiscriminate contact between humans and animals, and a refusal to think of the planet's biodiversity in terms of interrelations and a kind of connectedness that pertains to forms of life. In other features of the same genre, viruses circulate as a metaphor in expressions conveying contagious speed (such as to "go viral"). They are uttered without reflection on their consequence. Conversely, in many disease thrillers they are denounced as evil objects that valiant human subjects are assigned to destroy. *Contagion* is different: tracing the contagion to a zoonosis, that is, to a viral transmission from animals to humans, to a lack of hygiene, and by implication to ecological abuse, the film emphasizes the issues that literally plague a viral culture. It shows its failure to point to the complex origins of a virus and to overlooked problems concerning how we ought to inhabit the planet.

Viruses cannot simply be metaphorized or objectified. Pre-dating humans and much surpassing them in numbers, vital to life, viruses are ubiquitous. Expert voyagers, like passengers in international air carriers, said to travel in the atmosphere at high altitudes, they easily replicate, mutate, and jump from one species to another in their relentless push to find a welcoming host. Like bacteria, but of a simpler composition, they are *outside* of humans, swarming all around them, but also on and inside them, and even in the helical strands of DNA. Karin Moelling, a viral researcher, argues that viruses may not be only foes but also friends, indispensable for living creatures. Viruses, she writes, are part of our genome. The human being is an "ecosystem" characterized not by war but by a well-balanced coexistence and co-evolution with viruses: "The 'war' vocabulary has to be abandoned. Things only get dangerous when we destroy the balance" (2017, 6).

Such destruction seems to be driving many pandemics. As *Contagion* makes clear, given the rapid economic growth and expansion we are witnessing in every corner of the globe, many plant and animal habitats are disturbed, and effectively the turmoil leads to increased contact between viruses, animals, and humans. It has been amply documented through the study of various zoonoses that viruses are transmitted among different animals (camels, chimpanzees, pigs, birds, bats, insects, etc.) and humans. The circulation has less to do with the animals themselves than with how we humans go about inhabiting the earth.

The viral culture depicted in *Contagion* begs us to think closely about *ethology*, a term that Gilles Deleuze had adapted to describe Baruch Spinoza's *Ethics* (1988, 27). *Ethics* in Spinoza do not refer to moralizing or to a deontology, but to ways of being and inhabiting. For the Dutch philosopher, our ways of being are linked to a circulation of *affects*. The world is felt intimately, in and through our bodies, and not simply seen at a distance. Joyous or sad affects make organisms thrive or wither away. They increase or decrease an appetite for living. The focus on affects draws attention to the materiality of the body and of existence. Ethology, a term originally denoting the study of animal behavior mainly through objective human observation, has shifted in the currents of processual thinking to apply to humans and animals by way of a circulation of affects in multiple directions. Replacing a moral science, ethology refocuses the importance of ontology, a field of study it both complements and (in the best of ways) complicates. Ethology focuses on inhabiting and cohabiting with organisms, the consequences of which are extensive. Certain ways of inhabiting may be beneficial to some while destructive to others, to which old stories about biblical plagues already testify. Or as Deleuze writes: "Our place in nature seems to condemn us to bad encounters" (28). Some of the causes at the origin of those encounters can be dated from the eighteenth century and the Industrial Revolution in northern Europe that spurred colonial expansion and massive exploitation.

The rest, as it were, is history. Since the Enlightenment the world has witnessed widespread deforestation, intensive farming, the damming of rivers, thoughtless irrigation, rapid and often unplanned urbanization as well as a phenomenal increase in international trade and travel. Prior to COVID-19, cultural theorists, writers, and artists were still predominantly focused on ontological problems, while concurrently, like so many continental philosophers, scientists began looking closely at broader connections linking being and living or, say, ontology and ethology. For years they have warned of the risks of a viral pandemic. In 2018, WHO cautioned its public about the risk of a plague due to the multiplication of infections from viruses. They and many of their organizations are very aware of new major public health threats that are related to our ethos. Many have pledged their commitment to One Health, a concept that emphasizes the importance of human and animal welfare to sustainable development. Focused on areas where risks are elevated, specifically those with the most interaction of wildlife, livestock, and people in rapidly changing environments, their goal is to produce more sustainable conditions of life by identifying earlier, at local and regional levels alike, the possibility or nascent indications of disease. They also warn that given the shorter interval between epidemics and pandemics—SARS-CoV (2003), H1N1 (2009), MERS (2012), COVID-19 (2019), viral infections are here to stay. As recent mutants confirm, COVID-19 is not believed to be the terminus. Changes have to be introduced in the way humans inhabit the planet.

Yet, aside from scientists (whose research was severely curtailed and even suppressed under Trump), specialists in the humanities, world leaders, and the general public alike have paid little or no attention to the increased health risks in our culture that arise from viruses and their mutations. Moelling notes tersely, "viruses and bacteria behave opportunistically. They take advantage of unusual situations, of weaknesses of their hosts" (2017, 6–7). Viruses are like a pharmakon, they can be benevolent or malevolent. Affiliating a "viral culture" with social media, until now many of us have gone about business as usual. An exception might be Klaus Schwab, the founder of the Davos Economic Forum and author of *The Fourth Industrial Revolution* (2016). Following a theory of open systems, Schwab predicts (and even welcomes) the advent of what he calls tipping points. Included are new ways of inhabiting the world in 2025: the broad acceptance of driverless cars, robotics, 3D printing, and much more. His prognostics are written up in short, upbeat, enthusiastic chapters, praising new inventions and new ways of living, that end with thumbnail reviews of what they can offer and what they cannot. For Schwab, the benefits of the former, an easier and more convenient life, clearly outweigh the social and ecological drawbacks of the latter, mainly unemployment and environmental damage. To his description of this brave new world, he adds a caveat: these changes will come about barring an asteroid striking the planet, social upheavals...or a pandemic! While an asteroid always remains a remote possibility, social upheaval and pandemics have become real. Turmoil and confusion, now also related to COVID-19, have come to occur daily in many parts of the world. And the pandemic became a reality. In 2020–2021, COVID-19 put an abrupt (if perhaps only temporary) end to a frenzied global circulation of people—if not of goods—so well represented in *Contagion*. The pandemic may have altered some of Schwab's rosy visions. National borders were upended and carefully guarded, international travel curtailed, cities shut down, self-isolation turned into a fact of life, facial coverings made mandatory in public places. While the virus raged, restaurants, shops, small businesses, from hotdog vendors to beauticians to clothing stores, permanently closed; employees in the service sector risked their lives at work or lost their jobs. While many were pushed into poverty, profits soared at Microsoft, Amazon, Chewy, Google, Netflix, Pfizer, Zoom, and unexpected other corporations, construction companies, as well as bicycle manufacturers and outfitters, who all benefited from the new conditions of monadic (and hardly nomadic) existence and a sudden taste for the great outdoors.

How we reacted to the pandemic depended in part on where we found ourselves on the globe. In the United States, in an atmosphere marked by incessant buzz-feed, circulation of tweets, televisual journalism where conflicting reportages, sensationalist plots, and conspiracy theories abounded, the reaction was first of denial and even racial profiling: "This happens in China and Europe but not in the US!" The arrival of what Trump called an "alien" virus on American shores ignited fires of panic. The ensuing narratives continued to trace a line demarcating good and evil, concepts we have known for some time, that have gotten us beyond good and evil, as Nietzsche

(2006), Deleuze and Guattari (1987), and Rancière (1999) have shown time and again. Building on Spinoza's pronouncements on affects, Deleuze contended that "something bad" (but not "evil") is caught in a historical configuration or in a relation. Such is what we are told about the origins of COVID-19: allegedly spawned in the markets of live animals in Wuhan, China, the virus is said to have been transmitted by bats, by poachers of pangolins trafficking their scales to superstitious customers believing in their curative potential, and so forth. In knee-jerk reactions, with much hand-wringing, many exclaimed, if only the "others" acquired their foods in malls and shopping centers rather than live markets.[2]

The pandemic witnessed affluent members of the population seeking refuge in great numbers in second homes, from Cape Cod to the Hamptons and Destin, Florida, from Noirmoutier to Saint-Tropez, or the Costa Brava to the Baleares, where life went on unmasked, and where citizens were troubled only by the difficulties of country living—annoyances that, in turn, disgruntled and alienated local populations. By means of Zoom, Google Meet, Microsoft Teams, and so on, from the ease of remote location, many of the affluent continued to labor as usual, while others risked their lives in a system that, at least in the United States, was at first ill-prepared to cope with the pandemic. In *Contagion*, through their sacrifices, figures of the likes of Kate Winslet's character gain praise as national heroes. Less heard are the voices of those engaged in the service industry who live in cramped places from where they home-school their children (as Marion Cotillard's character discovers). The self-isolation of the affluent populations during the pandemic was in fact but a displacement in space from the way they inhabit the city in "normal" times. On the upper floors of glass towers, such as those of Hudson Yard in New York City or the exclusive apartments in the 8th and 16th arrondissements in Paris, residents of means were already self-isolating, although, perhaps, with more of their customary attributes and accompaniments (wine cellars, party-planners, dog groomers, and the like) (Conley 2020). Committee meetings on Zoom are spaces of congregation but also of segregation.

Perhaps not only the affluent but all of us who transferred our existence to "platforms" continued to live in a viral internet culture without reflecting on what it means to "break out" of our breakout rooms and protective enclosures. Long ago, in *Simulations*, Baudrillard (1983) wrote that we live in a script from a Hollywood horror film. Now, we live on the internet, on YouTube, and in social media, in scripts that both politicians and marketers aggressively and insidiously use with ever more sophisticated algorithms to construct and impose "lifestyles" upon entire populations. In addition to making people live in an atmosphere of capture, marketers exploit the sheer speed at which messages circulate to twist and insert them in areas far from the context of their origin or purpose. A general climate of misinformation prevails. It remains for all of us to ask ourselves how we can go beyond our ethos of living in a "viral" culture that we also help create. How can we combine our research and care or, simply put, how might we think of inhabiting otherwise by shifting toward an ethological way of thinking and living?

To exit the scenarios of control (many based on a logic of binary thinking), and to account for how we are enmeshed with the world, I would like to argue, in anticipation of a conclusion, as *Contagion* shows, that it is incumbent upon us *to care for the event* of encounter that led to the spread of the virus. For this, as suggested above, we have to think critically of our metaphoric use of "viral" and reflect on how viruses live and circulate in our biosphere. We can think of an approach that would not follow the colonizing model of a subject dominating an object or a simple division into good and evil. Clearly, the dominant ethos that came into existence with the onset of the Industrial Revolution is *not* working. Today, it is "nature" or "Gaia," the name of a force humans do not quite understand, that is reacting to environmental depredation. It is through our ways of inhabiting that we confront global health problems related to the coronavirus. That epidemics and pandemics occur at increasingly shorter intervals should be a warning about where a "viral" global culture, a culture that disturbs and destroys so many worlds, is headed. As Isabelle Stengers, Belgian feminist scientist-turned-philosopher has remarked, "when the city is extended into a cosmopolitan world, it is a world of which humans are citizens, everywhere at home, not a world where many worlds fit" (2018).

Scientists, including those at the Precision One Health Unit (POHU) in Geneva, or at the EcoHealth Alliance in New York, who study the global health of Human–Animal ecosystems, have argued that if our viral culture was quick to attribute the causes of COVID-19 to live animal markets, for the sake of the health of our ecosystem it is wiser to take account of a much larger entanglement of humans and animals. Live or wet markets exist in many parts of the world, notably North America where fish and seafood are sold. For many, the live market will continue to remain a source of sustenance. Western sensibilities do not have the same reaction to their own inhuman treatment, not only of fish and lobsters, but also of chickens, pigs, cows, and other animals farmed and raised for slaughter (Donna Haraway [2016]; Jacques Derrida [2008]; Nicholas Kristof [2014, 2020, 2021]). Live or wet markets become more dangerous when wild animals are added to the mix (bats, pangolins, monkeys, snakes, to mention just a few). Not only are these animals highly stressed and suffer terribly, they are also carriers of viruses that may not affect them but will precipitate disastrous effects when crossing over to humans. Yet, if markets are viral hotspots, they are but one element in the emergence of today's zoonoses that have given rise to epidemics and pandemics. In addition to clear-cutting that produces farmland of limited utility, mining for precious minerals required for the hardware of cell phones and computers augments human-animal contacts, mainly in tropical parts of the world with a higher density of wildlife. Mineral extraction contributes to direct and indirect global health risks. To limit them, we face the task of rethinking how we can go about reshaping the components of a viral culture.

Given the model set forward in *Contagion*, we would do well to consider the role humanities can play by drawing attention to how we inhabit our milieus and in what ways we can contribute to change reigning sensibilities.

But how? In an interview I conducted with Jacques Derrida in 1983 in the context of sexual difference, he responded with a query: why would you still want to call it that? He added that if the discourse of anatomy still dominates "modern discourse" such as psychoanalysis, the underlying disciplines of knowledge are no longer what they were. They go very, very fast (Conley 1995, 268). It is incumbent, he suggested, for the humanities to join in conversation with them: "What obscurantism there would be in ignoring current or future mutations!" (268). Varying on and fleshing out Derrida's observation at the onset of the twenty-first century, Stengers advocates a collaboration with what she no longer calls disciplines but *practices*. We can consider her query by transforming our otherwise enclosed, siloed disciplines into open-ended, collaborative, and collectively run endeavors. In her conversation with scientists, artists, proponents of social reform and others, she urges us to ask: "Is another world possible?" (Stengers 2011a). Practitioners have to experiment by working together; they must invent and compose a world of many worlds, but also be aware that habitation is always cohabitation. They are pragmatists of sorts who take care of the possible. In other words, they attend to what *can* happen and speculate on what *could* be. Just as we cannot separate humans from the world, we cannot separate disciplines.

Stengers is one of the few theorists today who are critical of the metaphor of growth and progress that subtend our viral culture and so often go unquestioned. Yet the pandemic has provided perhaps an opportunity for reflection. Despite—or in view of—death, misery, and hardship in our midst, in the abrupt halt of our daily routines, we may now be led to ask: do we need what we had or, rather, what is it that we really need? In his new book, *Covid-19: The Great Reset* (2020), co-authored with Thierry Malleret, Schwab himself revises his former utopian vision, although from a corporate point of view. Seen through the lens of the humanities, we can say that the pandemic hopefully leads us to reflect on our modes of existence, on how we inhabit and cohabit, and on how we might go about composing, collectively, a world in which space and time can be allotted for many other worlds.[3] However hypothetical or speculative a thought this may be, a different world could be envisioned, other sensibilities developed, and so also other modes of livelihood engaged and enacted. Such a world would not be reinvented from scratch. It is not a question of going "back" to some pristine world but, as Deleuze and Guattari already argued, of taking off from today's world, from what has been called our "capitalist ruins" (Tsing et al. 2017).

Thinking with Stengers in the way she is "thinking with Whitehead," we can countenance even more than an ethology, an etho-ecology that fosters the creation and implementation of local ways of inhabiting and exchanging that remain in touch with broader ecologies of the entire universe (Conley 2021).

As the pandemic has shown once again, the world is not just a background against which humans move and act (Serres 1996). Rather, it is a set of interdependent processes and resources capable of assemblages and combinations very different from the ones that currently exist. It is only in cohabitation and co-evolution with others—including viruses—that life forms can endure. Scientists tell us that the window or the sustained life of large mammals, humans included, is quite small. As in *Contagion*, the advent of vaccines to prevent the spread of COVID-19 attests to the necessity of collaborative effort. However, in the midst of the solitude and confinement in which we have been living during the pandemic, we are led to pause and reconsider the nature of a viral culture in the context of what Latour calls the general *incurie* [uncaring] of "techno-hype" (Latour and Leclercq 2016). We can ask what it does to inoculate us from paying attention to how entangled we are in different and often conflicting ecosystems. In addition to experimenting with life-saving vaccines, we have to speculate about other modes of existence. To avoid engaging in massive extermination, whether of humans or nonhumans, we can think of how to collaborate with those engaged in other practices—be it in the arts, architecture, urban policy, economics, and so forth—to underscore the need to compose another world and to bring about real transformation. Soon to be a half-century ago, at the height of postmodern atopia, in *The Three Ecologies*, Guattari pleaded for renewed attention to the subject, or to what he called subjectivation. He argued that those engaged in the above practices—to which we can add the sciences, the technosciences, economics—were responsible, for the creation of subjects that would connect with others and the world rhizomatically, indeed adventitiously and in many directions at once. Today, we can stress even more the need to compose assemblages in consort with an etho-ecology that focuses on cohabitation with humans, fauna, and flora. Scientists often point out that a massive, collective effort, both local and global, would roll back, even help to prevent further disasters and environmental collapse.

Ironically, it is in the business world, in the field of fashion, that some (reluctant) efforts are being made in this direction. Setting itself apart, the outfitter Patagonia argues responsibly for reusing old garments and even for making do with less. In an industry so plagued by waste and massive ecological abuse in addition to labor issues, under increasing consumer pressure, emphasis is shifting toward paying attention to a much-publicized "circular economy," that is, to a complete cycle from production to disposal. There is a faint impetus for quality over quantity and for making do with less. The attention of large companies (LVMH, Kering, Capri Holdings) is being seen to shift, toward three main issues: social justice, animal welfare, and issues of sustainability that they still fail to regroup under one banner. Though these global megaliths are dependent on growth for profit and care especially about customer retention, the shift nonetheless shows that they adapt under pressure from the changing sensibilities of younger customers. The question is: is it sufficient, is it happening fast enough, and what would this world look like?

First and foremost, it is contingent on a sliding of sensibilities and values already well under way. In conversation with other practices, the humanities can contribute by shifting their emphasis toward an etho-ecology in an ongoing experimentation that builds on its longstanding critical process, which relentlessly calls into question a four-century-old belief in binary thinking and in human control of the biosphere. We can take cues from French and Francophone theorists who have written on problems of inhabiting, from Marc Augé, Michel de Certeau, Gilles Deleuze, Vinciane Despret, Félix Guattari, Bruno Latour to writers, Marie Darrieussecq, stage directors, Frédérique Aït-Touati, or filmmakers, Jean-Luc Godard, Agnès Varda, and others. We will conclude, however, by varying one more time on Isabelle Stengers. She writes that an ecology of practices must also take into consideration what differentiates a practice from an idea. Practices cannot, any more than living beings, address a silent world or a docile substrate of convictions and interpretations (Stengers 2007). The practices for which she advocates are relational. They do not refer to an abstract authority for which they would be a local translation, but to a here-and-now that favors their possibility. To make perceptible for her is not to demonstrate a value, in the name of generalities that make one feel good such as "Peace" or "The Good" (Stengers 2015, 35). It is to produce an ethos that works and that makes all those involved live well and even thrive. It respects what Deborah Rose called "the shimmer of the world."[4]

With practices, the tiresome opposition of "good and evil" is supplanted by a differential scale of "good and bad," that in Soderbergh's *Contagion* withholds standard moral judgments while implying that we consider connections and relations to be a question of life and death. Bad connections literally poison organisms. The connection of a bat virus and a pig virus is not "evil," it is bad for humans who connect with it. It brings a plague! An ethos that leads to the inattentive razing of palm trees is bad for a local and global ecology. An etho-ecology is based on experimenting together, on attending to modes of existence, and on ways of affecting and being affected. It is based on the substitution of judgmental values with ethological values that are not argued but produced and performed (Stengers 2015, 58). We can shift our attention from "visions of the world" and metaphysical questions that claim to be valid for all, to local versions, to the singularities of what "matters" here and not somewhere else. We are called upon to take care of technical details that, as Stengers puts it, no one but the practitioners involved would deem worthy of interest, but which for them are the difference between success and failure.

The pandemic has shown that our viral culture is in need of a reset. To do so, we can turn our critical focus from a simple ontology toward an ethology and an etho-ecology that do away with divisions between social and natural dilemmas. Elizabeth Kolbert (2021) recently wrote that we have to control the very idea of control. To control our urge to control is to move toward other ways of experimenting and inhabiting the world based on ways of thinking that, at this point in time, in a world felt to be a vibrant continuum, are decidedly processual. The search for good modes of existence is always ongoing; it never reaches a final truth, and the outcome is always uncertain. To transform our culture of which COVID-19 is a symptom and a product, we are called upon to develop research as a form of care, to experiment with other ways of inhabiting at a time when our window of opportunity is shrinking. If the humanities are to collaborate with other practices and to experiment etho-ecologically with composing a world consisting of many worlds, they have little time to waste.

Acknowledgments

Conley, Verena. "Inhabiting a Viral Culture." *Substance* 51:1 (2022), 120–135. © 2022 Johns Hopkins University Press and SubStance, Inc. Reprinted with permission of Johns Hopkins University Press.

Notes

1 On Day 1, we learn that the Chinese chef extends his contaminated hand to Beth Emhoff (Gwyneth Paltrow), thereby transmitting the virus that will set off the pandemic. To inculpate the chef and make the woman the carrier of death can be seen as racist and misogynist. However, the real culprit in the film is the global company AIMM Alderson that carelessly razes a forest and displaces bats carrying the virus. "Contagion" refers to the pandemic, but also to a certain virulent capitalism that, capturing affects and destroying eco-systems, precipitates unwarranted catastrophes.

2 For a complementary and more sobering view of Wuhan, see Ai Wei Wei's remotely filmed documentary, *CoroNation* (2020). The opening aerial shot of over twenty rows of high-speed trains points by contrast to an ultramodern, expanding city.

3 A 200,000-ton containership with a carrying capacity of 20,000 shipping containers ran aground in the Suez Canal on March 23, 2021, while on route from Asia to Europe and the eastern United States. Pointing to the magnitude of etho-ecological challenges, the mishap of the Ever Given, chartered by the Evergreen (!) Marine Corporation, gave us another look into the extent of our consumer society.

4 In Chloé Zhao's *Nomadland* (United States, 2020), Swankie said she is ready to die after perceiving such a shimmer when she encounters a flock of swallows.

Works Cited/Consulted

Augé, Marc. 2008. *Non-Places*. Translated by John Howe. London: Verso.

Baudrillard, Jean. 1983. *Simulations*. Translated by Paul Foss, Paul Batman, and Philip Beitchman. New York: Semiotext(e).

Conley, Verena. 1995. "Voices." Interview with Jacques Derrida. In *Points...: Interviews, 1974–1994*, translated by Peggy Kamuf, edited by Elizabeth Weber, 156–170. Stanford: Stanford University Press.

Conley, Verena. 2013. *Spatial Ecologies: Urban Sites, State and World-Space in French Cultural Theory*. Liverpool: Liverpool University Press.

Conley, Verena. 2020. "Inhabiting Luxury Spaces." In *The Third Realm of Luxury: Connecting Real Places to Imaginary Spaces*, edited by John Armitage and Joanne Roberts, 67–84. London: Bloomsbury.
Conley, Verena. 2021. "A Plea for Etho-Ecology." In *Ethology: Claims and Limits of a Lost Discipline*, edited by Andrea Allerkamp and Martin Roussel, 83–91. Paderborn: Wilhelm Fink Verlag.
Deleuze, Gilles. 1988. *Spinoza, Practical Philosophy*. Translated by Robert Hurley. San Francisco: City Lights Books.
Deleuze, Gilles, and Félix Guattari. 1987. *A Thousand Plateaus*. Translated by Brian Massumi. Minneapolis: University of Minnesota Press.
Derrida, Jacques. 2008. *The Animal That Therefore I Am*. Translated by David Wills. New York: Fordham University Press.
Despret, Vinciane. 2016. *What Would Animals Say If We asked The Right Questions?* Translated by Brett Buchanan. Minneapolis: University of Minnesota Press.
Despret, Vinciane, et al. 2019. *Habiter le trouble avec Donna Haraway*. Bellevaux: Éditions Dehors.
Guattari, Félix. 1995. *Chaosmosis: An Ethico-Aesthetic Paradigm*. Translated by Paul Bains and Julian Pefanis. Bloomington: Indiana University Press.
Guattari, Félix. 2014. *The Three Ecologies*. Translated by Ian Pindar and Paul Sutton. London: Bloomsbury Academic.
Haraway, Donna. 2016. *Staying with the Trouble*. Durham: Duke University Press.
Kolbert, Elizabeth. 2021. *Under A White Sky*. New York: Crown Publishers.
Kristof, Nicholas. 2014. "Abusing Chickens we Eat." *New York Times*, December 3, 2014.
Kristof, Nicholas. 2020. "The Mistakes That Will Haunt Our Legacy." *New York Times*, July 11, 2020.
Kristof, Nicholas. 2021. "The Ugly Secrets Behind the Costco Chickens." February 26, 2021.
Latour, Bruno. 2001. *War of the Worlds: What About Peace?* Chicago: Prickly Press.
Latour, Bruno. 2015. "Sensitizing." In *Experience: Culture, Cognition and the Common Sense*, edited by Caroline Jones, 315–338. Cambridge: MIT Press.
Latour, Bruno. 2018. *Down to Earth: Politics in the New Climate Regime*. Translated by Catherine Porter. London: Verso.
Latour, Bruno, and Christophe Leclercq, eds. 2016. *Reset Modernity*. Cambridge: MIT Press.
Manning, Erin, and Brian Massumi. 2014. *Thought in the Act*. Minneapolis: University of Minnesota Press.
Massumi, Brian. 2015. *The Power at the End of the Economy*. Durham: Duke University Press.
Moelling, Karin. 2017. *Viruses: More Friends than Foes*. London: World Scientific Publishing.
Nietzsche, Friedrich. 2006. *The Essential Nietzsche: Beyond Good and Evil and The Genealogy of Morals*. Dover Publications.
Rancière, Jacques. 1999. *Disagreement*. Translated by Julie Rose. Minneapolis: University of Minnesota Press.
Schwab, Klaus. 2016. *The Fourth Industrial Revolution*. Geneva: Forum Publishing.
Schwab, Klaus, and Thierry Malleret. 2020. *Covid-19: The Great Reset*. Geneva: Forum Publishing.
Serres, Michel. 1996. *The Natural Contract*. Translated by Elizabeth MacArthur and William Paulson. Minneapolis: University of Michigan Press.
Stengers, Isabelle. 2007. *Cosmopolitics I*. Translated by Robert Bononno. Minneapolis: University of Minnesota Press.
Stengers, Isabelle. 2011a. *Thinking With Whitehead*. Translated by Michael Chase. Cambridge: Harvard University Press.

Stengers, Isabelle. 2011b. "The Care of the Possible." Interview with Erik Bordeleau. Translated by Kelly Ladd. *Scapegoat: Architecture | Landscape | Political Economy, Service* 1 (Autumn): 12–27.
Stengers, Isabelle. 2015. *In Catastrophic Times: Resisting the Coming Barbarism.* Translated by Andrew Geoffrey. London: Open Humanities Press.
Stengers, Isabelle. 2018. "The Challenge of Ontological Politics." In *A World of Many Worlds*, edited by Marisol Cadena and Mario Blaser, 83–111. Durham: Duke University Press.
Tsing, Anna. 2015. *The Mushroom at the End of the World: On the Possibility of Life in Capitalist Ruins.* Princeton: Princeton University Press.
Tsing, Anna, Nils Bubandt, et al. 2017. *The Arts of Living on a Damaged Planet.* Minneapolis: University of Minnesota Press.

Index

Pages in *italics* refer to figures and pages followed by "n" refer to notes.

For Product Safety Concerns and Information please contact our EU representative GPSR@taylorandfrancis.com
Taylor & Francis Verlag GmbH, Kaufingerstraße 24, 80331 München, Germany

www.ingramcontent.com/pod-product-compliance
Ingram Content Group UK Ltd.
Pitfield, Milton Keynes, MK11 3LW, UK
UKHW020006200926
475702UK00015B/2497
9781032741826